Shadows

Similar Triangles and Proportional Reasoning

Teacher's Guide

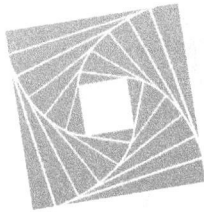

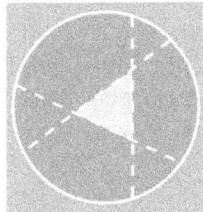

This material is based upon work supported by the National Science Foundation under award numbers ESI-9255262, ESI-0137805, and ESI-0627821. Any opinions, findings, and conclusions or recommendations expressed in this publication are those of the authors and do not necessarily reflect the views of the National Science Foundation.

Key Curriculum
1150 65th Street
Emeryville, California 94608
email: editorial@keypress.com
www.keycurriculum.com

First Edition Authors
Dan Fendel, Diane Resek, Lynne Alper, and Sherry Fraser

Contributors to the Second Edition
Sherry Fraser, Jean Klanica, Brian Lawler, Eric Robinson, Lew Romagnano, Rick Marks, Dan Brutlag, Alan Olds, Mike Bryant, Jeri P. Philbrick, Lori Green, Matt Bremer, Margaret DeArmond

Project Editors
Joan Lewis, Sharon Taylor

Consulting Editor
Mali Apple

Editorial Assistant
Juliana Tringali

Professional Reviewer
Rick Marks, Sonoma State University

Calculator Materials Editor
Christian Aviles-Scott

Math Checker
Christian Kearney

Production Director
Christine Osborne

Executive Editor
Josephine Noah

Textbook Product Manager
Tim Pope

Publisher
Steven Rasmussen

Contents

Blackline Masters

1-Inch Graph Paper Blackline Master
1/4-Inch Graph Paper Blackline Master
1-Centimeter Graph Paper Blackline Master
In-Class Assessment
Take-Home Assessment

Introduction

Shadows Unit Overview

Intent

This final unit of Year 1 asks the question, "How long is a shadow?" Students will bring their developing facility with patterns, functions, and algebra to bear on this question, as well as exploring important ideas in geometry and trigonometry.

Mathematics

The concept of similarity is the central theme of this unit. Through this concept, students explore the following important ideas from geometry and algebra.

Similarity and Congruence

- Developing intuitive ideas about the meaning of "same shape" and learning the formal definitions of *similar* and *congruent*

- Discovering the special properties of triangles in connection with similarity, as well as other features of triangles as special polygons

- Understanding the role of similarity in defining the trigonometric functions of sine, cosine, and tangent

Proportional Reasoning and the Algebra of Proportions

- Understanding the meaning of proportionality in connection with similarity

- Developing equations of proportionality from situations involving similar figures

- Understanding the role of proportionality in nongeometric situations

- Developing techniques for solving equations involving fractional expressions

Polygons and Angles

- Developing angle sum formulas for triangles and other polygons

- Discovering the properties of angles formed by a transversal across parallel lines

- Discovering the triangle inequality and investigating its extension to polygons

Logical Reasoning and Proof

- Working with the concept of *counterexample* in understanding the criteria for similarity

- Proving conjectures about vertical angles and polygon angle sums

- Understanding the role of the parallel postulate in proofs

Right Triangles and Trigonometry

- Learning standard terminology for triangles, including *hypotenuse, leg, opposite side,* and *adjacent side*

- Learning the right triangle definitions of *sine, cosine,* and *tangent*

- Using sine, cosine, and tangent to solve real-world problems

Experiments and Data Analysis

- Planning and carrying out controlled experiments

- Collecting and analyzing data

- Identifying key features in graphs of data

Mathematical Modeling

- Using a geometric diagram to represent a real-world situation

- Using scale drawings to solve problems

- Applying properties of similar triangles to real-world situations

- Exploring how models provide insight in a variety of situations

Progression

The unit begins with explorations in *What Is a Shadow?* of the variables that affect the length of a shadow. Students use the approach to gathering data via controlled experiments developed in *The Pit and the Pendulum.* The activities in *The Shape of It* and *Triangles Galore* develop important ideas about similarity and the geometry of triangles and other polygons. In *The Lamp Shadow* and *The Sun Shadow,* students solve the lamp shadow problem using side ratios in similar triangles and the sun shadow problem using similarity applied to right triangles—that is, trigonometry.

What Is a Shadow?

The Shape of It

Triangles Galore

The Lamp Shadow

The Sun Shadow

Pacing Guides

50-Minute Pacing Guide (27 days)

Day	Activity	In-Class Time Estimate
1	What Is a Shadow?	
	How Long Is a Shadow?	45
	Homework: *Experimenting with Shadows*	5
2	Discussion: *Experimenting with Shadows*	20
	The Shadow Model	25
	Introduce: *POW 13: Cutting the Pie*	5
	Homework: *Poetical Science*	0
3	Discussion: *Poetical Science*	10
	Shadow Data Gathering	40
	Homework: *An N-by-N Window*	0
4	Discussion: *An N-by-N Window*	15
	Working with Shadow Data	30
	Homework: *More About Windows*	5
5	Discussion: *More About Windows*	15
	The Shape of It	
	Draw the Same Shape	35
	Homework: *How to Shrink It?*	0
6	Discussion: *How to Shrink It?*	20
	The Statue of Liberty's Nose	30
	Homework: *Make It Similar*	0
7	Discussion: *Make It Similar*	15
	Presentations: *POW 13: Cutting the Pie*	20
	Introduce: *POW 14: Pool Pockets*	15
	Homework: *A Few Special Bounces*	0
8	Discussion: *A Few Special Bounces*	10
	Ins and Outs of Proportions	35
	Homework: *Similar Problems*	5

9	Discussion: *Similar Problems*	10
	Inventing Rules	40
	Homework: *Polygon Equations*	0
10	Discussion: *Polygon Equations*	10
	Triangles Galore	
	Triangles Versus Other Polygons	35
	Homework: *Angles and Counterexamples*	5
11	Discussion: *Angles and Counterexamples*	15
	Why Are Triangles Special?	35
	Homework: *More Similar Triangles*	0
12	Discussion: *More Similar Triangles*	10
	Are Angles Enough?	40
	Homework: *In Proportion*	0
13	Discussion: *In Proportion*	20
	What's Possible?	30
	Homework: *Very Special Triangles*	0
14	Discussion: *Very Special Triangles*	20
	Angle Observations	25
	Homework: *More About Angles*	5
15	Discussion: *More About Angles*	15
	Presentations: *POW 14: Pool Pockets*	20
	Introduce: *POW 15: Trying Triangles*	10
	Homework: *Inside Similarity*	5
16	Discussion: *Inside Similarity*	20
	A Parallel Proof	25
	Homework: *Angles, Angles, Angles*	5
17	Discussion: *Angles, Angles, Angles*	10
	The Lamp Shadow	
	Bouncing Light	30
	Homework: *Now You See It, Now You Don't*	10
18	Discussion: *Now You See It, Now You Don't*	10
	Mirror Magic	40
	Homework: *Mirror Madness*	0

19	Discussion: *Mirror Madness*	15
	A Shadow of a Doubt	25
	Homework: *To Measure a Tree*	10
20	Discussion: *To Measure a Tree*	20
	Presentations: *POW 15: Trying Triangles*	15
	Homework: *POW 16: Spiralaterals*	15
21	*More Triangles for Shadows*	25
	The Sun Shadow	
	The Sun Shadow Problem	25
	Homework: *Right Triangle Ratios*	0
22	Discussion: *Right Triangle Ratios*	15
	Sin, Cos, and Tan Revealed	10
	Homemade Trig Tables	25
	Homework: *Your Opposite Is My Adjacent*	0
23	Discussion: *Your Opposite Is My Adjacent*	15
	Homemade Trig Tables (continued)	10
	The Tree and the Pendulum	25
	Homework: *Sparky and the Dude*	0
24	Discussion: *Sparky and the Dude*	15
	A Bright, Sunny Day	25
	Beginning Portfolio Selection	10
25	*Beginning Portfolio Selection* (continued)	10
	Presentations: *POW 16: Spiralaterals*	20
	Homework: *Shadows Portfolio*	20
26	*In-Class Assessment*	40
	Homework: *Take-Home Assessment*	10
27	*Assessment* Discussion	30
	Unit Reflection	20

90-Minute Pacing Guide (17 days)

Day	Activity	In-Class Time Estimate
1	What Is a Shadow?	
	How Long Is a Shadow?	35
	Experimenting with Shadows	35
	The Shadow Model	20
	Homework: *Poetical Science*	0
2	Discussion: *Poetical Science*	10
	Shadow Data Gathering	40
	Working with Shadow Data	30
	Introduce: *POW 13: Cutting the Pie*	5
	Homework: *An N-by-N Window*	5
3	Discussion: *An N-by-N Window*	15
	More About Windows	35
	The Shape of It	
	Draw the Same Shape	40
	Homework: *How to Shrink It?*	0
4	Discussion: *How to Shrink It?*	20
	The Statue of Liberty's Nose	35
	Make It Similar	35
	Homework: *POW 13: Cutting the Pie* (continued)	0
5	Presentations: *POW 13: Cutting the Pie*	25
	Ins and Outs of Proportions	40
	Introduce: *POW 14: Pool Pockets*	15
	Homework: *A Few Special Bounces*	10
6	Discussion: *A Few Special Bounces*	10
	Similar Problems	35
	Inventing Rules	40
	Homework: *Polygon Equations*	5
7	Discussion: *Polygon Equations*	10

	Triangles Galore	
	Triangles Versus Other Polygons	25
	Angles and Counterexamples	30
	Why Are Triangles Special?	25
	Homework: *More Similar Triangles*	0
8	*Why Are Triangles Special?* (continued)	10
	Discussion: *More Similar Triangles*	15
	Are Angles Enough?	35
	What's Possible?	25
	Homework: *In Proportion*	5
9	Discussion: *In Proportion*	20
	Very Special Triangles	40
	Angle Observations	30
	Homework: *More About Angles*	0
10	Discussion: *More About Angles*	10
	Presentations: *POW 14: Pool Pockets*	20
	Introduce: *POW 15: Trying Triangles*	10
	Inside Similarity	30
	A Parallel Proof	20
	Homework: *Angles, Angles, Angles*	0
11	Discussion: *Angles, Angles, Angles*	10
	The Lamp Shadow	
	Bouncing Light	25
	Now You See It, Now You Don't	20
	Mirror Magic	35
	Homework: *Mirror Madness*	0
12	Discussion: *Mirror Madness*	20
	A Shadow of a Doubt	25
	To Measure a Tree	45
	Homework: *POW 15: Trying Triangles* (continued)	0
13	Presentations: *POW 15: Trying Triangles*	20
	Introduce: *POW 16: Spiralaterals*	15
	More Triangles for Shadows	30

	The Sun Shadow	
	The Sun Shadow Problem	25
	Homework: *Right Triangle Ratios*	0
14	Discussion: *Right Triangle Ratios*	15
	Sin, Cos, and Tan Revealed	15
	Homemade Trig Tables	35
	Your Opposite Is My Adjacent	25
	Homework: *POW 16: Spiralaterals* (continued)	0
15	*The Tree and the Pendulum*	25
	Sparky and the Dude	25
	A Bright, Sunny Day	25
	Beginning Portfolio Selection	15
16	*Beginning Portfolio Selection* (continued)	10
	Shadows Portfolio	30
	In-Class Assessment	40
	Homework: *Take-Home Assessment*	10
17	*Assessment* Discussion	30
	Presentations: *POW 16: Spiralaterals*	30
	Unit Reflection	30

Materials and Supplies

All IMP classrooms should have a set of standard supplies and equipment, and students are expected to have materials available for working at home on assignments and at school for classroom work. Lists of these standard supplies are included in the section "Materials and Supplies for the IMP Classroom" in *A Guide to IMP*. There is also a comprehensive list of materials for all units in Year 1.

Listed below are the supplies needed for this unit.

Shadows

- Flashlights (one or two per group)
- Cubes (about 10 per group)
- Tape measures (one per group)
- Metersticks and yardsticks
- Straws (about 10 per group)
- String (non-stretch; dental floss works well)
- Grid chart paper
- Scissors (one per group)
- Pipe cleaners
- Mirrors (one or two per group)
- Dry spaghetti or paper strips

More About Supplies

- Straws are a commonly used manipulative. Plastic straws work best and are available at grocery stores.

- Flashlights are used several times during the Shadows unit. Students could be asked to bring flashlights from home, but inexpensive flashlights are available at local home improvement stores, drug stores, and discount or dollar stores.

- Graph paper is a standard supply for IMP classrooms. Blackline masters of 1-Centimeter Graph Paper, ¼-Inch Graph Paper, and 1-Inch Graph Paper are provided so you can make copies and transparencies for your classroom. (You'll find links to these masters in "Materials and Supplies for Year 1" of the Year 1 guide and in the Unit Resources for each unit.)

Assessing Progress

Shadows concludes with two formal unit assessments. In addition, there are many opportunities for more informal, ongoing assessment throughout the unit. For more information about assessment and grading, including general information about the end-of-unit assessments and how to use them, see "Assessment and Grading" in *A Guide to IMP*.

End-of-Unit Assessments

Each unit concludes with in-class and take-home assessments. The in-class assessment is intentionally short so that time pressures will not affect student performance. Students may use graphing calculators and their notes from previous work when they take the assessments. You can download unit assessments from the *Shadows* Unit Resources.

Ongoing Assessment

Assessment is a component in providing the best possible ongoing instructional program for students. Ongoing assessment includes the daily work of determining how well students understand key ideas and what level of achievement they have attained in acquiring key skills.

Students' written and oral work provides many opportunities for teachers to gather this information. Here are some recommendations of written assignments and oral presentations to monitor especially carefully that will offer insight into student progress.

- *Shadow Data Gathering* and *Working with Shadow Data:* These activities, which ask students to set up and conduct controlled experiments (as in the unit *The Pit and the Pendulum*), will provide evidence of their understanding of the unit problems.

- *Similar Problems:* This assignment will provide evidence of students' ability to write and solve proportions derived from similar figures.

- *Angles and Counterexamples:* This activity will help you assess students' ability to create and solve linear equations derived from a geometric context and their developing understanding of similarity.

- *Angles, Angles, Angles:* This assignment will give you information on students' knowledge of facts about angles created by intersecting lines (including transversals of parallel lines) and interior angles of polygons.

- *Mirror Madness:* This activity will tell you whether students can use the reflective property of mirrors along with the concept of similarity to do indirect measurement.

- *A Shadow of a Doubt:* This activity will provide evidence about whether students understand the general solution to the lamp shadow problem.

- *The Tree and the Pendulum:* This assignment will illustrate students' ability to use trigonometry to do indirect measurement.

- *A Bright, Sunny Day:* This activity will provide evidence of students' understanding of the general solution to the sun shadow problem.

Supplemental Activities Overview

Shadows contains a variety of activities at the end of the student pages that you can use to supplement the regular unit material. These activities fall roughly into two categories.

- **Reinforcements** increase students' understanding of and comfort with concepts, techniques, and methods that are discussed in class and are central to the unit.

- **Extensions** allow students to explore ideas beyond those presented in the unit, including generalizations and abstractions of ideas.

The supplemental activities are presented in the teacher's guide and the student book in the approximate sequence in which you might use them. Below are specific recommendations about how each activity might work within the unit. You may wish to use some of these activities, especially the later ones, after the unit is completed.

Some Other Shadows (reinforcement) This is a very open-ended assignment in which students investigate other aspects of shadows than those considered in the unit. You can offer it to students any time after they have a clear idea of what the lamp shadow and sun shadow problems involve.

Investigation (extension) This activity, which provides a general framework in which students can do further experimental work with relationships and data, would be appropriate any time after the initial experimental work and perhaps be best after the discussion of *Working with Shadow Data*. You may want to have students work in pairs on this project and give them some time to brainstorm possible topics.

Cutting Through the Layers (extension) Students search for a function that depends on two variables, a challenge raised by the introductory activities of the unit. They gather and organize data from a geometric context and then look for patterns and relationships.

Explaining the Layers (extension) Students develop a formula for a generalization of the situation presented in *Cutting Through the Layers*.

Crates (extension) In this three-dimensional extension of the activities *An N-by-N Window* and *More About Windows,* students search for a function in three variables.

Instruct the Pro (reinforcement) Students create a complicated polygon and then write instructions for re-creating the figure using only a ruler and protractor. This activity supports their early work with angle measurement and the use of protractors.

Scale It! (reinforcement) Students investigate scaling by creating a scale model or a scale drawing, or speaking with someone who does these things, and then write a report connecting this open-ended investigation to their work in the unit.

The Golden Ratio (extension) Students explore the ratio that has been described as the most aesthetically pleasing for the dimensions of a rectangle. This activity would probably be meaningful to students once they have started thinking about ratios and is especially recommended for those who might be motivated by connections between mathematics and art.

From Top to Bottom (extension or reinforcement)   Students scale a pentagon to fit a standard sheet of paper oriented in first one direction and then the other and find ratios between the original drawings and their scale drawings.

Proportions Everywhere (extension or reinforcement)   Students use their work in *The Shape of It* to find the ratios among corresponding parts of similar figures. You may want to assign this after students work on the supplemental activity *From Top to Bottom.*

How Can They Not Be Similar? (reinforcement)   This activity continues students' investigation of criteria for similarity, encouraging them to look very carefully at the "corresponding parts" aspect of similarity. (It is possible to construct two pentagons that fit the given conditions. Though we are not aware of any example of two quadrilaterals that fit the conditions, no one we have consulted knows of a proof that this is impossible.)

*Rigidity Can Be Good (*extension)   Students investigate the significance of geometric rigidity in the fields of architecture and construction. This activity is appropriate after students have played with the materials in *Why Are Triangles Special?* and have seen that triangles form rigid structures.

Is It Sufficient? (reinforcement)   This geometric exploration, a continuation of ideas about logic and counterexamples, can be used after students have worked on *Equations with Angles* and *More Counterexamples* and *Why Are Triangles Special?* Students may be perplexed (justifiably) about what it means to say that "a side of one triangle is proportional to a side of the other triangle" (see Condition 2 of this activity). If questions arise, you might help students to articulate the idea that any single number is "proportional to" any other number, so this condition on sides is not truly meaningful.

Triangular Data (reinforcement)   In *Why Are Triangles Special?* students saw that three side lengths determine a triangle (the congruence property traditionally abbreviated SSS). This activity offers an opportunity for them to develop the other standard congruence conditions (abbreviated SAS, ASA, and AAS) and to see that two sides and a non-included angle do not necessarily determine a triangle.

What If They Kept Running? (extension)   Students apply concepts of proportionality in a context involving distances and rate of speed.

An Inside Proof (extension)   This activity asks students to prove two statements about a triangle and a line segment drawn through the triangle parallel to one of the triangle's sides, a situation discussed in connection with *More About Angles.*

Fit Them Together (reinforcement)   This activity prompts students to begin thinking about what happens to the area of a polygon when its dimensions are doubled. You might get them started with an illustration of the special case of squares.

What Is a Shadow?

Intent

In these activities, students familiarize themselves with the physical situation that creates shadows. They use data-collection and data-analysis skills (developed in *The Pit and the Pendulum*) and In-Out tables (introduced in *Patterns*) to identify and explore the effects of several variables on the length of a shadow.

Mathematics

The length of the shadow cast by an object near a lamp is a function of the height of the lamp, the height of the object, and the distance of the object from the lamp. When a shadow is created by sunlight, the length of the shadow is a function of the height of the object and the angle of elevation of the sun.

In *What Is a Shadow?,* students encounter these two situations and experiment with the "lamp shadow" problem in some detail. They build a mathematical model of the physical situation by making some assumptions and using plane figures in two dimensions.

Several activities ask students to use both In-Out tables and an analysis of the geometry of the situation to find functional relationships. Students come to understand that data analysis alone will not help them find the complicated three-variable function that will allow them to determine the length of a lamp shadow.

Progression

What Is a Shadow? begins informally, with students identifying and describing in qualitative terms the variables that affect the length of a lamp shadow. Then students undertake a series of controlled experiments to isolate the effects of each of these variables. Finally, they try to find several other functional relationships in geometric contexts. Following these activities, the unit moves away from the two-part unit problem—on what affects the lengths of lamp shadows and sun shadows—to develop important ideas of similarity and plane geometry.

How Long Is a Shadow?

Experimenting with Shadows

The Shadow Model

POW 13: Cutting the Pie

Poetical Science

Shadow Data Gathering

An *N*-by-*N* Window

Working with Shadow Data

More About Windows

How Long Is a Shadow?

Intent

In this activity, students begin thinking about what causes a shadow and what variables might affect the length of a shadow. They also design initial experiments to test their ideas.

Mathematics

This activity introduces students to the concepts of a "lamp shadow" (a shadow created by a nearby light source like a streetlamp) and a "sun shadow." Your shadow changes length as you walk away from a lamp, but does not change length as you walk away from the sun. These two situations frame students' work throughout this unit.

In this activity, students focus on the first situation. They consider the variables that might affect the length of a lamp shadow and—drawing on data-collection and analysis techniques developed in *The Pit and the Pendulum*—begin to plan experiments to test the affects of these variables.

Progression

Working in groups, students begin by considering situations that produce shadows and then identify variables that might affect the length of a lamp shadow. Finally, they devise plans for testing the effects of each of these variables.

Approximate Time

45 minutes

Classroom Organization

Groups, followed by whole-class discussion

Materials

Flashlight

Doing the Activity

Have students read the opening sections of the activity, and discuss what causes shadows. Then have them consider the questions in the section "What Is a Shadow?" You may want to have students begin with a discussion in groups and then bring the class together to share ideas.

Ask, **What causes shadows?** Students should express that shadows are created by the contrast between an area that receives light and an area where the light is blocked off (at least partially). You might ask one student to shine a flashlight on a wall and have another student interpose an object so that students see the outline of the object in shadow. Students can then move either the flashlight or the object to see how this changes the shadow.

In particular, focus on the issue of why some types of shadows change size when you move while others don't. Use students' responses to bring out the distinction between two types of shadows.

"Sun shadows" are caused by the sun (or moon), where the light source is, in practical terms, infinitely far away. If a person walks toward or away from the sun along a level surface, the length of the person's shadow along the ground does not change.

"Lamp shadows" are caused by such things as a lamp, flashlight, or streetlight. Changes in distance from the light source will affect the length of a lamp shadow.

Tell students that the central problem for this unit—exploring what affects the length of a shadow—has two parts. For most of this unit, students will investigate lamp shadows. They will return to the issue of sun shadows near the end of the unit.

Have students work in their groups for about 10 minutes to come up with a list of variables that they think might influence the length of a shadow. Ask that they be as specific as possible, particularly in describing measurements.

Then ask the class, **What variables did you list?** Compile and post a class list of potential variables on chart paper.

Next, have groups select a variable to work with and begin planning their lamp shadow experiments. For example, they might choose to use a streetlight, lamp, or flashlight as the light source and a person, tree, or so forth as the object.

Discussing and Debriefing the Activity

Have a few students share their group's plan for tonight's experiments. Students will individually carry out these experiments outside of class in the activity *Experimenting with Shadows.*

Key Questions

What causes shadows?

What variables did you list?

Experimenting with Shadows

Intent
Students conduct initial experiments to explore whether particular variables really affect the length of a shadow and, if so, how changing the values of those variables affects the shadow length.

Mathematics
Experimental design is one of the important mathematical ideas underlying this activity. These experiments will give students a general sense of the problem rather than reliable data for analysis. They will do more careful experiments in a couple of days, applying what they learn in this activity.

Progression
These experiments are meant to engage students with the idea of shadows and what causes them and to help them realize that more than one variable affects the length of a shadow.

Approximate Time
30 minutes for activity (at home)

20 minutes for discussion

Classroom Organization
Individuals, then groups, followed by whole-class discussion

Materials
Measurement tools (such as rulers or metersticks)

Flashlight or lamps

Objects (such as cubes) to create a shadow

Doing the Activity
Students will follow the plan devised by their groups in the activity *How Long Is a Shadow?*

Discussing and Debriefing the Activity
You might have students work for about five minutes in their groups to discuss the difficulties and the results of their shadow experiments and to formulate a summary statement of their findings. While they are doing this, identify students with good diagrams to prepare transparencies for display later.

Ask volunteers to present their groups' conclusions about their chosen variables. **What conclusions did your group reach from the experiments?** At this point, students will likely focus on qualitative conclusions rather than on detailed quantitative findings. For example, they may say "The taller the object, the longer the shadow," without giving a formula relating shadow length to height.

Record groups' summary statements along with the posted list of variables that was compiled in the previous activity.

Discuss the importance of performing careful experiments. Ask, **What difficulties did you have in carrying out your experiments?** Have students share concerns about the logistics of their experiments—materials used, measurement difficulties, and so forth.

Tell students that they will soon be doing more precise experiments, in which each group will again focus on a single variable, to gather numeric data that they might use to get a better understanding of how each variable affects shadow length.

Key Questions

What conclusions did your group reach from its experiments?

What difficulties did you have in carrying out your experiments?

The Shadow Model

Intent

This activity will help students to "mathematize" the lamp shadow situation. Students will develop a geometric diagram identifying the key variables that affect the length of a lamp shadow.

Mathematics

In this activity, students develop a **mathematical model** for a physical situation. In this case, the model is a representation of the key relationships in a three-dimensional, real-world situation that is reduced to a two-dimensional representation of overlapping right triangles. The focus in this work is on three variables: the height of the light source (L), the distance along the ground from the object to the light source (D), and the height of the object (H).

Progression

Students revise their experiment diagrams and label the variables. A class discussion then formulates one of the two main goals for the unit: finding a formula for the length of a lamp shadow in terms of the variables L, D, and H.

Approximate Time

25 minutes

Classroom Organization

Individuals, followed by whole-class discussion

Doing the Activity

The discussion that follows assumes that students are doing the activity by hand. If they have access to dynamic geometry software, however, they may want to create their shadow diagrams as sketches within the software rather than on paper. Students are asked to refine the diagrams of their shadow experiments to create a more precise mathematical model of the lamp shadow problem. To help focus them on the task, ask,

What exactly did you measure?

If you were to reduce this diagram to its simplest form, what would it look like?

Discussing and Debriefing the Activity

Have students work together to develop a class diagram. One aspect of developing the diagram is the elimination of details that are mathematically extraneous. For example, you might suggest that students represent the light by a point and the ground by a straight line. As students discuss the diagram, focus their attention on the lengths labeled *L, D,* and *H* below. Once the class develops such a diagram, post it for reference throughout the rest of the unit.

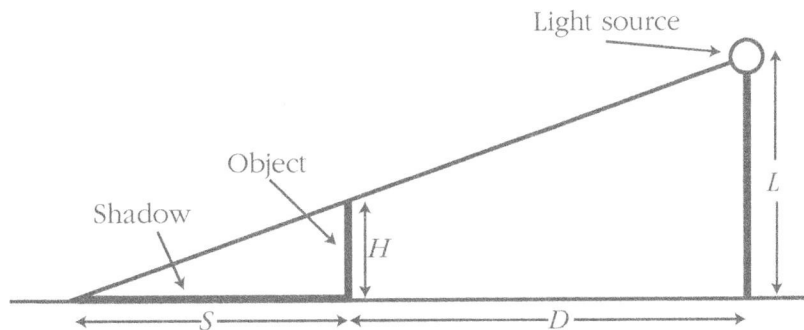

As needed, help students articulate precise verbal descriptions of each measurement in the diagram. For example, they might offer these descriptions:

- *L* is the distance from the light source to the ground.

- *D* is the distance along the ground from the light source to the object casting the shadow.

- *H* is the height of the object casting the shadow.

Introduce the symbols *L, D,* and *H* as abbreviations for these measurements. Also introduce the symbol *S* for the length of the shadow itself, measured along the ground. You may want to have the class look over the list of variables they generated initially and discuss any items on that list other than the variables *L, D,* and *H* defined here. Some of the variables may be relevant only to sun shadows; others (such as the brightness of the light) may be extraneous to the length of any shadow.

Using such a diagram involves some assumptions about the situation, and it may help to make these explicit. Ask students, **What assumptions are you making by using this diagram?** For example, the diagram assumes that

- the shadow is caused by a single light source

- the ground is level

- the object casting the shadow is vertical

You may want to post a list of these assumptions, perhaps leaving room for others to be added later.

Tell students that this diagram is a **mathematical model** for the shadow problem. Emphasize that the term refers to the use of a mathematical or abstract description of a real-world situation. Mention that such models generally involve simplifications or assumptions like those just discussed, so any conclusions that are based on a model should be tested in the real situation if possible.

Now ask, **How can you state the unit problem using these variables?** You will probably want to post the following, which is the first of the two-part unit goal:

Unit goal: To find a formula expressing S, the length of a lamp shadow, in terms of the variables L, D, and H.

Ask students if they can use $f(x)$ function notation to express the type of formula they will be looking for. **How can you state the goal using $f(x)$ notation?** As needed, help them develop a generic function equation, such as $S = f(L, S, H)$. Explain that this notation is essentially shorthand suggesting that there could be an In-Out table with three inputs (L, D, and H) and one output (S). In this notation, f represents the rule for the table.

Keep in mind that function notation is fairly new to students and often difficult for them to use. Also, this is likely to be the first time they are seeing function notation used with more than one independent variable. Don't worry if some students are not comfortable with this formulation of the problem. They will be working with function notation throughout the curriculum, and this exposure is just one opportunity to get them used to it.

Optional: More on Sun Shadows

Although the main focus of the unit is the lamp shadow problem, you may want to also talk about other shadow issues. Students will return to the sun shadow problem later in the unit, so you can discuss the diagram for the sun shadow problem now or to wait until it comes up later.

If the sun or another distant object were used as the light source, the problem changes character, because L and D cannot be treated as variables in the same way as is done for lamp shadows. If possible, get students to talk about the experience of walking along and seeing their shadow moving along with them, without changing length.

As will be discussed later, it makes sense in the sun shadow problem to focus on the angle from the light source to the object (shown as θ below) as one of the main variables. That is, students will be looking for a function of the form $S = f(H, \theta)$. You may want to introduce the Greek letter θ (theta) and explain that it is commonly used in mathematics to represent angles.

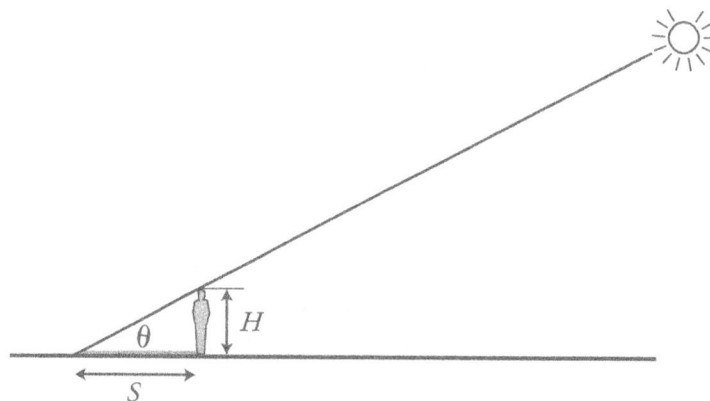

Even in the lamp shadow situation, one could use the angle as one of the parameters that affects the length of the shadow. However, the angle is not independent of *L, D,* and *H*. That is, it is impossible to keep *L, D,* and *H* fixed and change the angle. However, if the idea of an angle as a variable arises, you might acknowledge that one could use this as one of the basic parameters and explain that students will come back to look at this variable near the end of the unit when they examine sun shadows.

Key Questions

What exactly did you measure?

If you were to reduce this diagram to its simplest form, what would it look like?

What assumptions are you making by using this diagram?

How can you state the unit problem using these variables?

How can you state the goal using $f(x)$ notation?

Supplemental Activity

Some Other Shadows (reinforcement) investigates shadows that are cast along a surface that is not horizontal.

POW 13: Cutting the Pie

Intent

Students analyze the numeric pattern in a geometric problem and explain the pattern in terms of geometry. The activity gives students additional practice with collecting data, organizing information, and looking for patterns.

Mathematics

In *Patterns,* students had many opportunities to search for and generalize patterns in In-Out tables. This POW, which involves a fairly complex pattern analysis of a geometric problem, provides another. Because of the nonlinear nature of the relationship, finding a general formula relating the *In* values to the *Out* values in the table will probably be difficult for most students. They might have more success finding a recursive relationship for successive *Out* values. Understanding why that formula fits the geometry of the situation is another dimension of this POW.

Progression

Introduce this POW about a week before the write-ups are due.

Approximate Time

5 minutes for introduction

1 to 3 hours for activity (at home)

20 minutes for presentations and discussion

Classroom Organization

Individuals

Doing the Activity

Make sure students understand what they are to do. You may want to start the In-Out table as a class.

Discussing and Debriefing the Activity

Have presenters share their work. Students will most likely have noticed a sequential pattern in their data. It is more important that students be able to describe the pattern, and connect that pattern to the geometry of the problem, than that they develop a closed formula to represent it.

In the maximum case, how many new pieces does the *n*th cut make?

Students will probably see from the table that the nth output is *n* more than the previous output. If not, they may have incorrect entries, perhaps because they didn't find the maximum number of pieces possible. You may want to go over several cases to be sure everyone agrees on the maximums.

Probably at least one presenter will show how to use the pattern to find the maximum number of pieces for 10 cuts (Question 2b). If this does not come up in the presentations, ask the class about this.

Why does the pattern in the table hold?

You may want to ask students what ideas they have about why the *n*th output in the table is *n* more than the previous output. This question involves two issues, which students may or may not address.

- Why the *n*th cut cannot create more than *n* new pieces. If students are interested in examining this issue, you might have them start with a diagram with three cuts and slowly draw a fourth cut to see how this generates new pieces. This cut first creates a new piece by splitting the piece in which it begins and then, as it crosses each existing cut, splits the new piece it enters.

- Why it is possible to draw the lines so that *n* new pieces are created by the *n*th cut. This issue is more difficult. The key idea is to show that it is possible to draw the cuts so that each new cut intersects all previous cuts. Just making the cuts nonparallel to previous cuts is not enough, because the intersections need to be within the circle representing the pie.

Question 3 asks students to find a rule for the In-Out table. In one sense, they have already done so by describing the pattern by which the entries change. There are several other ways to find a rule, including the following. If these approaches do not all come up in the presentations, you may want to at least hint at the possibility of each one.

- As a summation expression: To see how to write the *n*th output as a sum, one can think of the difference from one entry to the next as the number of new pieces created by the *n*th cut. The total number of new pieces created by all *n* cuts is $1 + 2 + ... + n$. Because there is one piece (the whole pie) before any cuts are made, the nth entry is $1 + (1 + 2 + ... + n)$. If students get an expression like this, you might ask whether they can rewrite it using formal summation notation (for example, $1 + \sum_{t=1}^{n} t$).

- As a closed formula: Students may have experience with the sum $1 + 2 + ... + n$ that will allow them to develop a closed formula for it, that is, a formula without "..." or summation notation.

- As a recursive equation: Recursive notation provides a way to formalize the process of getting a table entry from the previous entry (or, more generally, from several preceding entries). You might suggest using subscript notation such as a_n to represent the maximum number of pieces for *n* cuts. If this is the first experience students have had with this type of notation, you might ask them, for instance, for the value of a_4 or a_{10}. Then ask if they can describe the rule using this notation. To be more concrete, start by asking how to use this notation to express a_{100}. It may help if you get students to state in words how they could find the 100th entry from the one before and then move toward the equation $a_{100} = a_{99} + 100$. Once students get this, they should be able come up with the general equation $a_n = a_{n-1} + n$, or the equivalent.

Review that such an equation is called a recursive formula. Point out that a recursive formula needs a "starting place," which in this problem is probably $a_1 = 2$ (because one cut creates a maximum of two pieces) or $a_0 = 1$ (because there is one piece before there are any cuts).

Key Questions

In the maximum case, how many new pieces does the nth cut make?

Why does the pattern in the table hold?

Poetical Science

Intent

In this activity, students have the opportunity to consider whether mathematics is an art or a science or both and to reflect on their own use of imagination to understand mathematics.

Mathematics

Albert Einstein once said, "I am enough of an artist to draw freely upon my imagination. Imagination is more important than knowledge. Knowledge is limited. Imagination encircles the world." There is an enduring debate about the nature of mathematics and whether it is an art, a science, or both. The general public perceives mathematics as the clearest example of a logical, structured activity that is the opposite of art, while mathematicians speak of the beauty, elegance, and creativity of their arguments and results.

Progression

Students read a brief biography that focuses on the nature of the scientific personality and then write about their own mathematics experiences.

Approximate Time

20 minutes for activity (at home or in class)

10 minutes for discussion

Classroom Organization

Whole class

Doing the Activity

If you have time, have students read the biography aloud in class. You may want to have a conversation about what students are being asked to do.

Discussing and Debriefing the Activity

You might have several students share their ideas about a time when they used their imagination in this class. You might also foster a discussion about what occupations require a great deal of imagination and why.

Shadow Data Gathering

Intent

In this activity and *Working with Shadow Data,* students become physically involved in the central unit problem, extending their explorations in *Experimenting with Shadows* to become more quantitative. They tackle the lamp shadow problem using an approach similar to that used in the previous unit, *The Pit and the Pendulum.*

Mathematics

Working with the key elements of experimental design and measurement precision, students devise and conduct controlled experiments to test the effects of each of the three variables in the lamp shadow situation on the length of the shadow. Groups record and use fixed values for two of the variables and then experiment to see how the shadow's length changes as they alter the variable they have chosen to investigate.

Progression

Each group tests one variable and posts their results. Groups then review each other's work to get a more complete picture of the effects of each of the three key variables on the length of a lamp shadow.

Approximate Time

40 minutes

Classroom Organization

Groups

Materials

Flashlights (1 per group)

Cubes (5–10 per group)

Metersticks or yardsticks (1 per group)

Grid chart paper

Doing the Activity

Remind students of their work in *Experimenting with Shadows* and *The Shadow Model,* and explain that they will now develop more precise descriptions of the effects of the three key variables on the length of a lamp shadow.

Each group will gather data relating the shadow length, S, to one of the other three variables—$L, D,$ or H—and will make an In-Out table of their data. Let each group choose one of the three variables—$L, D,$ or H—to experiment with, with at least one group (or, if possible, at least two groups) working with each variable. Emphasize the need to keep the other parameters fixed.

It will be helpful to tape over part of the flashlight opening to focus the light more directly on the object casting the shadow. It will also help if groups place their materials on grid paper with the same units as the cubes, so they can mark their measurements directly on the grid paper.

As groups work, they may ask about the level of accuracy needed in their measurements. You can let them work out their own ideas for dealing with this issue.

Groups should prepare their results as reports on chart paper to be posted for other groups to review. Reports should include an In-Out table with the variable for the *In* identified clearly as *L, D,* or *H*. They should also note the values used for the variables that were held fixed.

Discussing and Debriefing the Activity

Give students time to examine all of the posted group reports. When students examine the results, they may notice that, for at least two of the variables, no simple rule can be derived to explain the information in the In-Out table.

An *N*-by-*N* Window

Intent

Students search for a numeric formula to describe a new geometric situation, either by creating and examining an In-Out table or by analyzing the geometry of the situation.

Mathematics

This activity and *More About Windows* offer students practice in using patterns in In-Out tables or the geometry of the situations themselves to find functional relationships between variables. As with *POW 13: Cutting the Pie,* the nonlinear relationship between the variables in this activity makes finding a rule connecting *In* to *Out* using just the table data a challenge.

Progression

Students work on the activity individually prior to a whole-class discussion.

Approximate Time

25 minutes for activity (at home or in class)

15 minutes for discussion

Classroom Organization

Individuals, followed by whole-class discussion

Doing the Activity

This activity requires little or no introduction.

Discussing and Debriefing the Activity

You might have students share their formulas in groups before holding a class discussion.

Some students will have compiled an In-Out table like the one below. Students who created such a table may have been able to figure out a rule based on the data.

Length of side of window	Total length of wood needed
1	4
2	12
3	24
4	40

Some students may have taken a more analytical approach. For example, they may have noticed that there are $N + 1$ rows of horizontal wood strips, with N feet in each row, and an equal number of vertical wood strips.

If students used only the In-Out table to find a rule, ask whether they can see why that rule must work in all cases. This may lead them to the more analytical approach.

If students found different algebraic expressions that fit the data or describe the situation, such as $2N(N + 1)$ and $2N(N + 2)$, ask, **Can both of these expressions be correct? What would that mean?**

You might suggest that groups try to reconcile the different formulas, looking for reasons why they give the same results. They might graph the formulas as functions on the graphing calculator to see whether they produce the same graph.

Key Question

Can both of these expressions be correct? What would that mean?

Working with Shadow Data

Intent

Groups will use the data they collected in *Shadow Data Gathering,* and the data gathered by at least one other group, to develop some general observations about how the chosen variable affects shadow length.

Mathematics

Groups will search for formulas to fit the data from their shadow experiments. Only one of the variables is linearly related to shadow length, so finding specific rules is not the primary goal of this activity. One of the ultimate goals of the unit, however, is to find a function relating shadow length, S, to the three independent variables L, D, and H.

Progression

Groups look for relationships between shadow length and the key shadow variables, prepare reports summarizing their findings, and then share their results.

Approximate Time

30 minutes

Classroom Organization

Groups

Materials

Posted reports from *Shadow Data Gathering*

Doing the Activity

Each group is to look for a relationship between shadow length and their chosen variable, based on the information in their In-Out tables.

When groups finish working with their own data, they can examine the posted data from other groups. They should work with data sets that study each of the other variables before examining another set focused on the variable that they investigated.

Groups are not expected to find equations connecting shadow length and the given variable. (Except for the case of the variable D, they would be unlikely to be able to do so.) Rather, the goal is to restate and describe more precisely the general qualitative relationships found in *Experimenting with Shadows*. Here are the types of statements they might make:

- As L increases, shadow length decreases. As L decreases, the shadow gets longer.

- As H increases, shadow length increases, but the shadow grows much faster than the height. As H gets close to L, the shadow length gets very large.
- As D increases, shadow length increases. The relationship between shadow length and D seems linear (which it is).

Discussing and Debriefing the Activity

Each group should prepare a written report summarizing its work. The report should include the results from *Shadow Data Gathering* and *Working with Shadow Data*—both for the group's own data and for the data of other groups.

You might have each group report their findings. **What relationships did your group find? How do those relationships show up in the graphs?** Also ask for intuitive explanations of these general relationships, based on the situation. **What in the situation might lead to that relationship?**

You might have the class compare the graphs from groups that investigated the same variable. Students may note that the general shape of the graphs is the same. Bring out that even if the groups used the same scales, the graphs will probably differ, because each group chose different values for the fixed variables.

Even though students were not asked to find formulas from their data, they may feel frustrated that they could not do so. They may see the shadow problem as analogous to their work on pendulums in the last unit and think they should be able to get a formula based on experimental results. If so, you might ask, **How is the shadow problem different from the pendulum problem?** You might ask what variables students considered in the pendulum problem and which turned out to be important, and then ask the same questions about the shadow problem. Help students see that here they are dealing with three variables that affect the length of the shadow. Another important distinction is that in T*he Pit and the Pendulum,* they were mainly interested in a specific situation, the 30-foot pendulum, while in this unit the goal is to find a general formula.

Ask students where they are now in terms of the central shadow problem. They should see that an approach based purely on gathering data is probably inadequate. Explain that in order to develop complete formulas relating S to L, D, and H, and to understand the relationships among these variables, students will need to understand some geometry. They will now spend several weeks studying geometry concepts that, later, will help them more precisely address the unit problem.

Key Questions

What relationships did your group find? How do those relationships show up in the graphs?

What in the situation might lead to that relationship?

How is the shadow problem different from the pendulum problem?

Supplemental Activity

Investigation (extension) provides a general framework in which students can do further experimental work with relationships and data.

More About Windows

Intent

This activity poses a more general "windows" situation in which students look for a pattern and an algebraic rule for two inputs.

Mathematics

As a follow-up to *An N-by-N Window,* this activity asks students to generalize their work to a rectangular *M-by-N* window frame. It is even less likely than before that an In-Out table will provide enough information for students to find a rule to represent the situation, making it even more important for them to examine the geometry of the situation.

Progression

Students work on this activity individually and then share their results with the class.

Approximate Time

20 minutes for activity (at home or in class)

15 minutes for discussion

Classroom Organization

Individuals, followed by whole-class discussion

Doing the Activity

This activity requires little or no introduction.

Discussing and Debriefing the Activity

Ask volunteers to share their ideas and approaches. As with *An N-by-N Window,* there are both experimental and analytic approaches to finding a rule.

Some students may have experimented with examples, made an In-Out table for their results with two inputs, *M* and *N,* and then looked for a rule to describe the relationship between the *Out,* the amount of wood strip needed, and the two *Ins.* The formula is not one that jumps out of the numbers easily, so students who work solely from a table may not have found a formula.

Other students may have developed a formula by analyzing the geometry of the situation.

Encourage students to use these two approaches to reinforce each other. For example, if a student developed a formula by analyzing the geometry, have that student test the formula with examples. If a student gathered data from examples and then found a formula, urge that student to look for an explanation of that formula in the context of the situation.

Supplemental Activities

Cutting Through the Layers (extension) presents another situation in which two inputs generate the output.

Explaining the Layers (extension) asks students to develop a formula for a generalization of the situation presented in *Cutting Through the Layers*.

Crates (extension) extends the ideas in *An N-by-N Window* to a three-dimensional context.

The Shape of It

Intent

After having done some initial exploration into the two "shadows" questions that drive this unit, students begin to develop some fundamental ideas about similarity of plane figures. The activities in *The Shape of It* and *Triangles Galore* lay the mathematical foundation for students to be able to solve the lamp shadow problem.

Mathematics

When are two plane figures the same shape? In the first part of *The Shape of It,* this somewhat informal question is answered in a formal way: two plane figures are the same shape if they are **similar**—that is, if their corresponding angles are equal in measure and their corresponding sides are proportional in length. In other words, one of the figures is a scaled-up version of the other. Students explore some of the consequences of this definition and, through this, develop some meaningful algebraic techniques for solving proportions.

Progression

The Shape of It begins with activities that formalize the concept of similarity and ends with activities that develop algebraic techniques for finding missing side lengths in pairs of similar figures. In addition, students will present their solutions to the first POW of the unit and begin work on the second.

Draw the Same Shape

How to Shrink It?

The Statue of Liberty's Nose

Make It Similar

POW 14: Pool Pockets

A Few Special Bounces

Ins and Outs of Proportion

Similar Problems

Inventing Rules

Polygon Equations

Draw the Same Shape

Intent

In this first activity of *The Shape of It,* which is devoted to developing basic ideas of similarity, students use concrete situations to develop an intuitive idea of the meaning of "same shape."

Mathematics

In this introductory activity, which sets the stage for eventually defining what it means for two plane figures to be similar, students are encouraged to reflect on what it means for two figures to be the same shape. Does orientation matter? Are two figures the same shape if they are simply the same *type* of figure (for example, two rectangles)? The house in the activity has a set of length and angle measurements. Students investigate which measurements change as they enlarge the house.

Progression

Students work on the activity individually and then discuss their work as a class.

Approximate Time

35 minutes

Classroom Organization

Individuals, followed by whole-class discussion

Doing the Activity

Be sure students recall how to use a protractor. They may also need a review of angles and of measuring polygon angles.

Discussing and Debriefing the Activity

Question 1, in which students draw a simple picture on grid paper and then draw one exactly like it, only larger, probably doesn't need discussion.

For Question 2, ask two or three students to draw their figures, labeling the measurements for each side and angle, and let the class discuss whether the figures are the same shape as Renata's house. At present, you can leave the question unresolved, perhaps explaining that, for now, students might disagree about what "same shape" means, but that they will be working toward the formal definition that mathematicians use.

Similarly, for Question 3, there will probably be some differences of opinion. Some students may say that "same shape" includes "facing the same way," so they will not consider the pair of rectangles in Question 3a or the pair of triangles in Question 3d to be the same shape. On the other hand, some may consider that all rectangular figures, for example, are "the same shape" (namely, they are all rectangles).

For Question 3c, some students may consider the triangles to be the same shape even though the height-to-width ratios are different. Some may even consider the pentagon and hexagon of Question 3b to be the same shape (because they are both polygons), even though the number of sides differs.

No conclusions need to come out of this discussion, but help students to define the issues that they agree and disagree about. You might post the following, for example, as questions to be resolved: **Is the mirror image of something considered the same shape? Does changing the size of something change its shape?**

Highlight comments that hint at the idea of ratio or scale, as this is one of the foundations of the formal definition of *similar*. For example, a student might say about the triangles in Question 3c, "The triangle on the right is taller so it should also be wider, but it isn't, so it isn't the same shape," which suggests that an increase in one dimension ought to be matched by an increase in another dimension.

Key Questions

Is the mirror image of something considered the same shape?

Does changing the size of something change its shape?

Supplemental Activity

Instruct the Pro (reinforcement) offers students additional practice with basic ideas about angle measurement and the use of protractors.

How to Shrink It?

Intent

This activity uses one of the situations posed in *Draw the Same Shape* to help students refine their sense of what it means for polygons to be considered the same shape. It also reinforces understanding of length and angle measurement, especially protractor use.

Mathematics

Students continue to develop ideas about "same shape" by examining three proposed methods of making a smaller version of a drawing: (1) by subtracting a fixed amount from each side length, (2) by dividing each angle by a fixed amount, and (3) by dividing each side length by a fixed amount.

Progression

Students work on their own to evaluate the three methods and then share their results with the class.

Approximate Time

20 minutes for activity (at home or in class)

20 minutes for discussion

Classroom Organization

Individuals, followed by whole-class discussion

Doing the Activity

Some students may find this activity redundant in light of the previous discussion. Emphasize that, so far, they have no formal definition of "same shape," so there are no obvious answers.

You can also challenge these students by drawing a vertical or horizontal line segment and asking them to make a drawing, using this segment as a particular side of a house, with the same shape as Renata's house.

Discussing and Debriefing the Activity

Begin by asking whether anyone can explain why each method will or will not work, and use the responses to gauge how much precision to press for. For example, if a student says, "Lily's method won't work because the angles aren't the same as Renata's," ask why the angles need to be the same.

Be alert for language that is leading up to the definition of similar to help move students toward articulation of this concept.

Key Questions

Why will Lola's method work or not work? What about Lily's? What about Lulu's?

The Statue of Liberty's Nose

Intent

Students use the context of the human body to explore the mathematical concepts of *similar, congruent,* and *corresponding parts* and relate these ideas to their earlier work and the unit problem.

Mathematics

This activity is designed to build an understanding of the formal definition of what it means for two figures to be the "same shape." Two plane figures are **similar** if (1) their corresponding angles are equal in measure and (2) their corresponding sides are proportional in length. There are two ways to recognize this proportionality: (1) pairs of sides within one figure must be in the same ratio as the corresponding pairs of sides within the other figure, and (2) each pair of corresponding sides must be in the same ratio as each of the other pairs of corresponding sides.

Progression

Students work in pairs to estimate the lengths of parts of the Statue of Liberty by "scaling up" from their own body measurements. As a class, they then relate this work to earlier activities and to the unit problem. The class discussion will focus on developing formal definitions of *similar* and *congruent*.

Approximate Time

30 minutes

Classroom Organization

Pairs, followed by whole-class discussion

Materials

String

Doing the Activity

This activity requires little or no introduction.

Discussing and Debriefing the Activity

Ask volunteers to present their answers to Question 1 and examples for Question 2. Students will presumably have used the idea of proportionality, although they might not use that word. For example, a student might say, "The Statue of Liberty's nose is about 25 times as long as my nose, so the statue's arm should be about 25 times as long as my arm."

There are two ways to use proportionality in Question 1. One approach is to find the nose-to-nose ratio for the two figures, statue and person, and apply that ratio to arms. The second approach is to find the arm-to-nose ratio for a person and apply that ratio to the statue.

The key idea connecting the Statue of Liberty problem to the "same shape" house problem (Question 2 of *Draw the Same Shape*) is that the statue is the same general shape as a person. If the language of ratio and proportion doesn't emerge in the discussion of Questions 1 and 2, try to elicit it in the discussion of Questions 3 and 4, introducing it yourself if needed.

For example, if a student uses the arm-to-nose approach described above, summarize this analysis by saying, "You saw that the ratio between the length of a person's arm and the length of a person's nose is about 15 to 1, so you wanted this ratio to be the same for the statue." Explain that when two ratios are the same, we say that the numbers involved are **proportional**. In addition, a statement that two ratios are the same is called a **proportion**.

Also use the term **corresponding parts** in the discussion. For example, point out that the statue's nose and a person's nose are corresponding parts of the two objects. Then ask, If the nose of the person and the nose of the statue are corresponding parts, what would be corresponding parts in the house problem? For instance, the floor of one house and the floor of another would be corresponding parts of two house diagrams.

Try to get students to combine the terms *ratio, corresponding parts,* and *same shape* in a summary statement something like this:

When objects are the same shape, the ratio of the lengths of one pair of corresponding parts is the same as the ratio of the lengths of any other pair of corresponding parts.

Ask whether students can state this principle using the word proportional. Work with them to come up with a statement like this:

When objects are the same shape, corresponding parts are proportional in length.

It will be helpful if students can achieve this level of clarity in a context as concrete as the Statue of Liberty problem so that they can use the language in the more abstract context of similar polygons.

For Question 5, let students share ideas on the connections between the activity and the unit problem.

Tell students that in mathematics, the term **similar** refers to objects that are the same shape. Emphasize that mathematicians are quite specific when they refer to *same shape* and that the definition you are giving today applies specifically to polygons. It may help to state the definition using the terms **ratio** and **proportional** so students see the connection between them. For example,

Two polygons (with the same number of sides) are similar if their corresponding angles are equal in size and the lengths of their corresponding sides are proportional (that is, the ratios of the lengths of corresponding sides are equal)

Post the definition prominently in the classroom. Emphasize that both conditions must hold in order for two polygons to be considered similar. (Technically, one should say that the angles are congruent, or equal in size or measurement, rather than equal.)

Next, introduce the symbol ~ for similarity. Illustrate it with an example such as the following, reading the symbols aloud by stating, "Triangle *PQR* is similar to triangle *STU*."

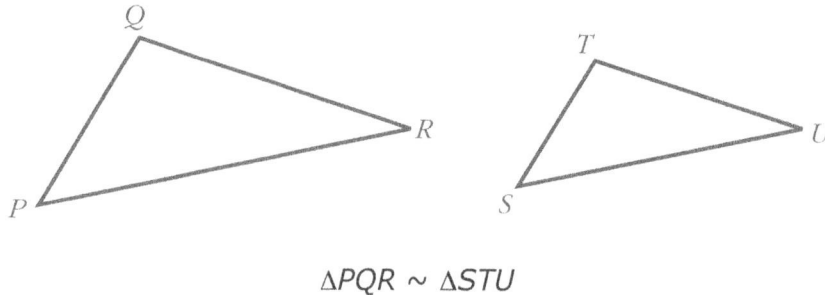

$$\triangle PQR \sim \triangle STU$$

Point out that the notation $\triangle PQR \sim \triangle STU$ indicates not only that the triangles are similar, but also how the vertices correspond (*P* to *S, Q* to *T,* and *R* to *U*). Students should not write $\triangle PQR \sim \triangle SUT$, for example, because that suggests that *Q* and *U* are corresponding vertices.

With the formal definition of *similar* established, ask students to look back at Question 3 of *Draw the Same Shape* and reach consensus about whether the figures in each pair are similar.

Students should see that the rectangles in Question 3a seem to be similar. Bring out that the corresponding sides here are not oriented the same way; the vertical sides of the first rectangle correspond to the horizontal sides of the second rectangle. Be sure students recognize that not every pair of rectangles is similar and that to show these two are similar, one should measure the lengths or otherwise verify that the second is simply a 90° rotation of the first.

Students should recognize that the triangles in Question 3d are similar (this again requires measuring). The figures in Question 3b cannot be similar because they have different numbers of sides (without the same number of sides, one can't set up corresponding sides and angles, so the polygons can't be similar). The triangles in Question 3c are not similar because, for example, the corresponding "top angles" are not equal.

Question 3a should also bring up the fact that similar figures do not have to be different sizes. Tell students that similar figures in which corresponding sides are equal in length—so that the ratio of corresponding sides is equal to 1—are called **congruent**. The two rectangles in Question 3a are congruent, and any polygon is congruent to itself.

Finally, Question 3a should raise the issue of the *orientation* of a figure. Although this may play a role in students' intuitive concept of "same shape," it is not part of the mathematical definition of similarity. Rotating a figure does not change its shape.

You can use Question 3d to clarify the meaning of **corresponding parts**. For instance, point to a side of one triangle and ask what the corresponding side is in the other triangle.

Although the figures in Question 3c are not similar, they can help clarify the concepts of proportionality and corresponding parts. Ask students which parts might be considered corresponding. They will presumably mention the two bases and the two pairs of sides. Then have students measure and find the ratio of each pair of corresponding parts of the two triangles. They should see that for one pair (the bases), the ratio is 1, but for the other two pairs, the ratio is something else.

Key Question

If the nose of the person and the nose of the statue are corresponding parts, what would be corresponding parts in the house problem?

Supplemental Activities

Scale It! (reinforcement) is an open-ended activity involving scale drawings.

The Golden Ratio (extension) explores the ratio that has been described as the most aesthetically pleasing for the dimensions of a rectangle. This activity is especially recommended for students who might be motivated by connections between mathematics and art.

Make It Similar

Intent

Students apply the definition of similar to determine which sides of two similar triangles are corresponding.

Mathematics

If you know that two triangles are similar, and if you know the three side lengths of one of the triangles, additional information is needed to find the side lengths of the other triangle. In this activity, students are given one side length of the new triangle, which leads to three possible sets of proportions and, thus, to three sets of answers. Students set up and solve these proportions using informal methods. More formal methods for solving proportionality equations are developed in upcoming activities.

Progression

Students work on this activity individually and then share their results in a class discussion.

Approximate Time

15 minutes for activity (at home or in class)

15 minutes for discussion

Classroom Organization

Individuals, followed by whole-class discussion

Doing the Activity

This activity requires little or no introduction.

Discussing and Debriefing the Activity

Discussion of this activity provides a good opportunity to clarify the concept of *corresponding sides.*

You might begin by asking, **How many solutions did you find?** Before getting into the details of any one solution, ask students to explain why there is more than one.

It may help to refer to the diagrams of the two triangles, perhaps drawing them side by side. Then ask for volunteers to describe their arithmetic in detail. **Exactly what did you do to find the lengths of the other sides?**

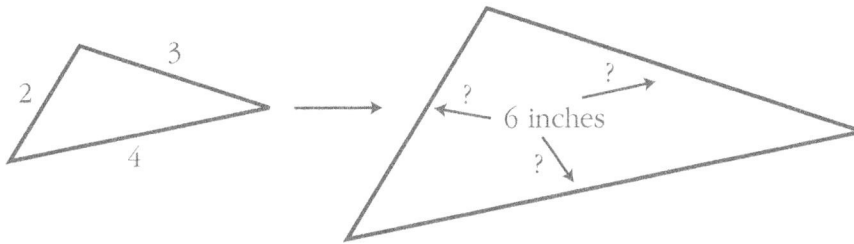

If possible, get students to articulate the need to find the ratio of the sides from one triangle to the other. One objective of this activity is for students to learn to set up such problems as proportions.

For example, in the case in which the 6-inch side of the second triangle is the longest side, the reasoning may be expressed as follows.

If the 6-inch side of the second triangle corresponds to the 4-inch side of the first triangle, the ratio of corresponding sides is 6 to 4. The same ratio applies to the other pairs of sides. To find the side length of the second triangle corresponding to the 2-inch side of the first triangle, solve the equation $\frac{x}{2} = \frac{6}{4}$.

If needed, identify an equation such as $\frac{x}{2} = \frac{6}{4}$ as a **proportion**. This proportion states that two fractions—that is, two **ratios**—are equal.

There are many ways to set up the proportions; encourage students to present alternatives. **Are there other ways to set up a proportion for this problem?**

Key Questions

How many solutions did you find?

Exactly what did you do to find the lengths of the other sides?

Are there other ways to set up a proportion for this problem?

Supplemental Activity

From Top to Bottom (extension or reinforcement) asks students to scale a pentagon to fit a standard sheet of paper oriented in first one direction and then the other.

POW 14: Pool Pockets

Intent

In this second POW of the unit, students look for patterns and generalizations in data collected for a situation that depends on two variables. They are encouraged to organize their data in a variety of ways as they investigate the problem.

Mathematics

In this activity, students explore the paths a ball will follow on an imaginary pool table in which the only pockets are those in the four corners. They investigate what happens to the ball—the number of bounces it takes off the walls of the table, whether it will land in a pocket and, if so, after how many bounces and in which pocket—and how these data depend on the height and width of the table. This open-ended problem provides a geometric context for gathering data, generating patterns, and generalizing relationships.

Progression

After some initial exploration, students work on the activity outside of class. In the activity *A Few Special Bounces,* they will explore a few special cases that will help them generalize this larger problem.

Approximate Time

15 minutes for introduction

1 to 3 hours for activity (at home)

20 minutes for presentations

Classroom Organization

Individuals

Doing the Activity

After students read the problem, facilitate a discussion to make sure they understand what is expected of them. Some may already have experience with the trajectory of a pool ball based on the angle at which it hits the side of the table; others may be unfamiliar with the context.

One general idea to encourage is organizing the individual cases in ways that might lead to insights. For example, students might look at tables that all have a common height (as done in the activity *A Few Special Bounces*).

Another fruitful idea is to consider the relationship between height and width. For example, students might consider cases in which the width is a multiple of the height. They might also try organizing cases according to which pocket the ball eventually lands in. For example, they might examine whether the cases in which the ball lands in the lower-right pocket have anything in common.

Discussing and Debriefing the Activity

Have the three presenters describe their results. When the presentations are over, ask whether anyone has other conclusions or other questions that may be worthy of investigation.

Key Question

What other conclusions did you find? What other questions did you investigate?

A Few Special Bounces

Intent

This activity will help students understand the situation they are investigating in *POW 14: Pool Pockets.*

Mathematics

In this activity, students are given a constraint for *POW 14: Pool Pockets* to explore. They consider all tables with a height of exactly two units.

Progression

Students work on the activity individually and share their results in groups.

Approximate Time

25 minutes for activity (at home or in class)

10 minutes for group discussion

Classroom Organization

Individuals, then groups

Doing the Activity

This activity requires little or no introduction. By starting with a special case, students will gain insight into how to approach and organize their work on the open-ended POW.

Discussing and Debriefing the Activity

Have students share their findings about this special case in their groups. Perhaps as a minimum, confirm that they found that the ball hit a pocket in every case.

Remind students that they are to consider other questions in the POW, in addition to which pocket the ball lands in. For example, they might explore the number of bounces needed for tables of different sizes.

Key Question

Did the ball always hit a pocket?

Ins and Outs of Proportion

Intent

This activity helps set the stage for the study of ratios within right triangles, which form the basis of trigonometry.

Mathematics

The activity builds indirectly on students' work with scale drawings and makes use of the algebraic skills students have been developing. Two principles of proportion for two similar triangles are identified: (1) the ratio of any pair of sides in one triangle is equal to the ratio of the corresponding pair of sides in the other triangle, and (2) the ratio of a side of one triangle and the corresponding side of the other triangle is the same no matter which side of the first triangle is used.

Progression

The questions in this activity lead students through the following sequence: (1) investigate ratios for a given pair of similar triangles, (2) create another pair of similar triangles and repeat, (3) create a pair of nonsimilar triangles and repeat, and (4) summarize results and conclusions using two similar triangles with sides labeled with letters. Students then share what they learn in a class discussion.

Approximate Time

35 minutes

Classroom Organization

Individuals, followed by whole-class discussion

Doing the Activity

The language in this activity may be difficult for students. Have volunteers read the activity out loud, and make sure everyone understands which sides of the triangles correspond before moving to ratios within a triangle.

Discussing and Debriefing the Activity

You might begin by having two or three students share their general results for Question 4. They should have concluded that if two triangles are similar, then a ratio of sides within one triangle is equal to the ratio of the corresponding sides in the second triangle.

In other words, the assumption of similarity—that the ratios $\frac{r}{x}$, $\frac{s}{y}$, and $\frac{t}{z}$ are equal—leads, for example, to the conclusion that the ratios $\frac{r}{s}$ and $\frac{x}{y}$ are equal.

Build as much as possible on students' earlier work with the algebra of proportions to bring out the connection between the geometry and algebra in this situation. In particular, discuss the idea that a triangle of a given shape has side lengths in a three-part ratio and that any similar triangle has the same ratio. For example, if one triangle has sides of lengths 5, 7, and 9, then the sides of any similar triangle must be in the ratio 5:7:9. (Though the word ratio is usually used for only two numbers, it can also be used in this way.)

Supplemental Activity

Proportions Everywhere (extension or reinforcement) is a fairly structured activity exploring proportionality in similar figures.

Similar Problems

Intent

This activity will help students formalize their work with the proportionality of corresponding sides of similar figures.

Mathematics

In this activity, students practice setting up proportions and solving equations. With the assumption of similarity, and knowing how the lengths of one pair of corresponding sides relate (thereby establishing the scale factor), students can write and solve equations to find missing side lengths in pairs of similar polygons.

Progression

Students work on the activity individually and then share their solution approaches in groups and with the whole class.

Approximate Time

5 minutes for introduction

15 minutes for activity (at home or in class)

15 minutes for discussion

Classroom Organization

Individuals, then groups, followed by whole-class discussion

Doing the Activity

When introducing the activity, you may want to talk about why measuring the diagrams will probably not give the correct answers.

Discussing and Debriefing the Activity

You might give students a few minutes to compare their equations and solutions in their groups and then have students or groups present individual problems to the class.

This is the first time in the unit that students have been asked to set up and solve formal equations for proportions. Ask presenters, How did you go about setting up the equations?

There is more than one way to set up an appropriate equation. For example, all three of the following equations can be used to express Question 1. The third equation equates ratios within the triangles rather than between the triangles.

$$\frac{x}{8} = \frac{15}{5} \qquad \frac{5}{15} = \frac{8}{x} \qquad \frac{5}{8} = \frac{15}{x}$$

Ask, **How did you solve the proportion equations?** Focus on the various solution methods. Many students will probably still primarily use intuitive methods, including trial and error.

However a solution is found, emphasize the importance of verifying that the suggested solution satisfies the equation by finding the value of each ratio. **How can you check that your solution is correct?**

If anyone suggests cross multiplying to solve a proportion, acknowledge the correctness of this technique without getting into why it works. Cross multiplication will arise in the discussion of the next activity, *Inventing Rules.*

Key Questions

How did you go about setting up the equations?

How did you solve the proportion equations?

How can you check that your solution is correct?

Inventing Rules

Intent

Students further develop their techniques for finding the missing values in proportions. They also articulate their own ideas about solving proportion equations.

Mathematics

The mathematical goal of this activity is to develop techniques for solving proportions. Students are given several proportions and are asked to find the missing values. Some students will propose cross multiplying as a method. This activity—along with *Ins and Outs of Proportions, Similar Problems,* and *Polygon Equations*—is designed, in part, to help students understand why cross multiplying and other manipulative techniques work.

Progression

Students work on their own to solve equations involving proportions and relate the proportions to similar triangles. The class then discusses cross multiplication and other techniques for solving proportion equations.

Approximate Time

40 minutes

Classroom Organization

Individuals, then groups, followed by whole-class discussion

Doing the Activity

Have students read the activity. You may want to have them, in groups or as a class, talk about ways they go about solving equations.

Discussing and Debriefing the Activity

You might assign one question to each group for presentation. Instruct group members to compare strategies and decide which method the class will understand the best.

Students may come up with some interesting explanations. Here are some samples of what you might hear.

- Question 1: "A fraction is like division, so this is saying $x + 5 = 7$. The answer is 35."

- Question 3: "The denominator of the first fraction is twice as big as the denominator of the second fraction, so the numerator must be twice as big also."

- Question 5: "Because $\frac{4}{6}$ is the same as $\frac{2}{3}$, $\frac{x+1}{3}$ has to equal $\frac{2}{3}$. This means that $x + 1$ is 2, so $x = 1$."

- Question 8: "The value for x has to be between 9 and 16. I tried all the numbers and 12 worked."

Students may find Question 7 difficult, especially if their conceptual understanding of fractions is weak. You may want to suggest a similar-looking but simpler problem such as $\frac{8}{x} = 4$ for comparison.

Questions 4 and 6 will likely be the most challenging. If the method of cross multiplying comes up, encourage students to offer some justification for the method. **Why do you think cross multiplying works?** If they can't explain why it works, ask them to think about why it might work.

The following two methods of explaining cross multiplication might be helpful. We use Question 4, $\frac{x}{7} = \frac{5}{3}$, to illustrate. First, ask students to express each fraction using a common denominator.

$$\frac{3x}{21} = \frac{35}{21}$$

Since the fractions have the same denominator, the numerators must be equal, so $3x = 35$. Or suggest that students multiply both sides of the equation by the common denominator, 21, and simplify, which also gives $3x = 35$. Multiplying both sides of an equation to get an equivalent equation should make intuitive sense to them.

In either case, ask, **Where did the 3x come from? Where did the 35 come from?** Help students to see that the two sides of the final equation could have been obtained by multiplying each numerator by the "opposite" denominator.

You can also use known pairs of equivalent fractions to verify that cross multiplying gives equal products.

You may also want to help students go through one of these methods in a more symbolic form, beginning with a general proportion, such as $\frac{a}{b} = \frac{c}{d}$, and using the common denominator, bd, to lead to the equation $ad = bc$. This general form may make it clearer where the final equation comes from.

Discuss Questions 9 and 10 as a class. Ask at least two students to draw the similar triangles for each of the questions.

Key Questions

Why do you think cross multiplying works?

Where did the $3x$ come from? Where did the 35 come from?

Polygon Equations

Intent

This culminating activity—following *Ins and Outs of Proportions, Similar Problems,* and *Inventing Rules*—continues students' work with developing and solving equations that connect corresponding sides of similar figures.

Mathematics

This activity will remind students of the tasks in *Similar Problems*. In this case, students are given similar polygons and are asked to set up equations to find the lengths of the sides labeled using variables. Some of the variables have a constant added to or subtracted from them, making the algebra slightly more difficult. Students also find any remaining lengths of the polygons using the information they have. They now have some new tools and techniques to bring to the task.

Progression

Students work on the activity individually and then discuss their work as a class.

Approximate Time

5 minutes for introduction

15 minutes for activity (at home or in class)

10 minutes for discussion

Classroom Organization

Individuals, followed by whole-class discussion

Doing the Activity

You may want to point out that students are to find the lengths of all the sides, not just those labeled. For example, in Question 1, they need to find the third side of the second triangle, which they should be able to do once they have found the value of t.

Discussing and Debriefing the Activity

This activity entails additional steps beyond setting up and solving equations, because even after finding the values of the variables, students will have to do additional work to find all the missing lengths.

Have students share their solution methods for each pair of polygons. Question 3 will likely lead to the equation $\dfrac{t}{t+1} = \dfrac{4}{6}$, although other equations are possible. Talk about how students might solve this equation, which has the variable in both the numerator and the denominator.

Triangles Galore

Intent

In *Triangles Galore,* students continue build an understanding of similarity and other important geometric relationships needed to solve the unit problem. In these activities, the focus is on relationships within and between triangles.

Mathematics

Triangles Galore further explores the concept of similarity, with a focus on similar triangles. Students have learned the definition of **similar**; now they will learn that, because of the rigidity of a triangle, it is sufficient to satisfy only parts of this definition to determine whether two triangles are similar. They will also explore several relationships among the parts of a triangle, such as the **triangle inequality** and the sum of the interior angles. For the special case of right triangles, students will be introduced to related terminology and will explore connections among the sides and angles in these polygons. Finally, they will analyze relationships among the angles formed by a **transversal** through **parallel lines.** Throughout the activities, students will write and solve equations.

Progression

Triangles Galore begins by asking students to contrast similarity in triangles with similarity in other polygons. Students then explore specific facts about triangles, their side lengths and measures, and the special case of right triangles. Next, they investigate relationships related to parallel lines. In addition, students will present their results for the second POW of the unit and begin work on the third.

Triangles Versus Other Polygons

Angles and Counterexamples

Why Are Triangles Special?

More Similar Triangles

Are Angles Enough?

In Proportion

What's Possible?

Very Special Triangles

Angle Observations

More About Angles

POW 15: Trying Triangles

Inside Similarity

A Parallel Proof

Angles, Angles, Angles

Triangles Versus Other Polygons

Intent

This activity presents conditional statements about triangles and asks whether the statements are true of other polygons. Their work will help students understand why triangles are special, especially in relation to similarity.

Mathematics

Are triangles different from other polygons? In this activity, students examine a series of statements that are true for similar triangles but not true for other polygons. For example, the fact that two angles of one triangle are equal in measure to two angles of another triangle is sufficient to guarantee similarity. In contrast, it is possible to construct two quadrilaterals in which two angles in one are equal to two angles in the other but that are not similar.

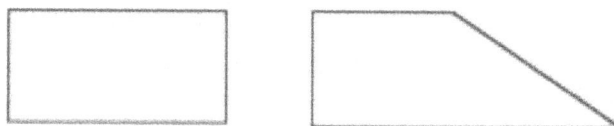

Students are asked to find counterexamples to prove that the statements are not true for polygons. They strengthen their understanding of basic mathematical logic through review of the term **counterexample** and introduction of the terms **hypothesis** and conclusion to understand "if, then" conditional statements.

Progression

Students work on the activity in pairs and share ideas in a class discussion.

Approximate Time

35 minutes

Classroom Organization

Pairs, followed by whole-class discussion

Doing the Activity

Review the meaning of the terms hypothesis and conclusion. Then introduce the activity by asking for a volunteer to answer this question: Who can state the definition of similar for polygons? Students should be able to give a statement like this:

Two polygons are similar if their corresponding angles are equal and their corresponding sides are proportional.

Tell students that in this activity, they will investigate the two conditions in this definition: equal corresponding angles and proportional corresponding sides.

Next, review the concept of a **counterexample** and its relationship to "if, then" statements. Students encountered this term in the first unit, *Patterns* (primarily in connection with consecutive sums), and they may have seen the "if, then" form that many mathematical generalizations take.

To review what a counterexample is, offer a statement like the following and ask whether it's true:

"All odd numbers are primes."

When students have put forth that the statement is false, ask for a specific number that shows it is false. When you get such a number, ask what that number is called in this context. If necessary, remind students of the term *counterexample*. Then ask, What exactly makes that number a counterexample? Students should be able to say something like, "The number 9 is a counterexample because it's odd but not prime."

The statement used above, "All odd numbers are primes," is not stated in "if, then" language. Restate it as something like, "If a number is odd, then that number is prime."

Finally, introduce the terms **hypothesis** and **conclusion** for the "if" and "then" parts of the statement, and ask students to use this language to explain what a counterexample is. As needed, help them to see that a counterexample fits the hypothesis (such as 9 for the "if, then" statement above) but does not fit the conclusion; it makes the hypothesis true and makes the conclusion false.

Have students explore the activity in pairs.

Discussing and Debriefing the Activity

Ask presenters how they know that a given statement is true for triangles and how they know that they have a counterexample for polygons. Make sure they verify both that the polygon fits the hypothesis and that it fails to fit the conclusion.

Bring out that not finding a counterexample is not the same as none existing. In other words, the fact that students didn't find a counterexample does not mean they have proved that the statement is true.

Key Questions

Who can state the definition of *similar* for polygons?

What exactly makes that number a counterexample?

Supplemental Activity

How Can They Not Be Similar? (reinforcement) continues students' investigation of criteria for similarity, encouraging them to look very carefully at the "corresponding parts" aspect of similarity. (It is possible to construct two pentagons that fit the conditions in the problem. Though we are not aware of any example of two quadrilaterals that fit the conditions, no one we have consulted knows of a proof that this is impossible.)

Angles and Counterexamples

Intent

Students practice writing and solving equations within the context of finding the measures of polygon angles. Then they look for more counterexamples for "if, then" statements that will help them distinguish triangles from other polygons.

Mathematics

This activity gives students more practice solving simple linear equations derived from a geometric context—in this case, the interior angles of polygons. In Patterns, students learned that the sum of the angles of an n-sided polygon is $180(n - 2)$; in the activities in *The Shape of It,* they will prove that this is true. In this activity, they will use this relationship to write and solve equations to determine the angle measures of several polygons.

Students will also once again search for counterexamples for several statements about triangles and other polygons. They should recognize that both conditions in the definition of **similar** are needed for identifying counterexamples, especially those concerning polygons other than triangles.

Progression

Students do this activity individually, check results in their groups, and then discuss their results as a class.

Approximate Time

5 minutes for introduction

20 minutes for activity (at home or in class)

15 minutes for discussion

Classroom Organization

Individuals, then groups, followed by whole-class discussion

Doing the Activity

Before students begin work on this activity, you may have to review the angle sum formula, which was developed in the *Patterns* activities *Degree Discovery* and *Polygon Angles.*

Discussing and Debriefing the Activity

Have students check their work on Questions 1–4 with their group members, and then have groups present their drawings and explanations for Questions 5–7 to the class.

Question 5 is straightforward. For Question 6, students might draw triangles *ABC* and *DEF* where angles *A* and *D* are the same size and the length of side *AB* is twice the length of side *DE* and *BC* is twice *EF,* but angle *C* is obtuse and angle *F* is acute.

Ask presenters, How do you know that you have a counterexample? Make sure they verify both that the polygons fit the hypothesis and fail to fit the conclusion.

For Question 7, students will probably believe that there are no counterexamples, but may be unsure how to articulate this. They should realize that if there really are no counterexamples, the statement is true. Bring out that not finding a counterexample is not the same as none existing. In other words, the fact that they didn't find a counterexample does not mean they have proved that the statement is true. Assure students that the statement in Question 7 is true, and let them talk about what makes them think so. They may be able to give an intuitive explanation based on symmetry.

Key Question

How do you know that you have a counterexample?

Why Are Triangles Special?

Intent

The focus in *Triangles Galore* is on triangles as a special figure, and in this activity this idea is addressed explicitly. Students investigate specific polygons to gain insight into why triangles are special with regard to similarity.

Mathematics

Triangles are special polygons, in part, because similarity implies that corresponding sides are proportional and proportional corresponding sides implies similarity. This is not true, in general, for other plane figures. For example, a rectangle and a nonrectangular parallelogram can have corresponding sides proportional but not be the same shape. In each figure below, the long sides are twice the length of the short sides.

In this activity, students learn through experimentation that a triangle with a given set of side lengths is "rigid." So, two triangles with the same side lengths are **congruent**, or the same shape and the same size. They also learn that for a given set of side lengths for figures with more than three sides, many different shapes (actually, an infinite number) are possible.

Progression

Groups explore the three questions in this activity—experimenting with quadrilaterals, polygons with more than four sides, and then with triangles—and share their findings in a class discussion.

Approximate Time

35 minutes

Classroom Organization

Groups, followed by whole-class discussion

Materials

Materials for constructing triangles, such as straws (10 per group), dental floss or string (several feet per group), and scissors

Doing the Activity

Once students have read the activity, offer them materials such as straws, scissors, and dental floss to build physical models of the polygons. They can cut the straws to various lengths to represent the sides of a quadrilateral and then string dental floss through the pieces so that the straws form a flexible sequence of line segments. The ends of the floss can then be tied together.

Discussing and Debriefing the Activity

Have students share their discoveries. They should recognize that, in the case of polygons with more than three sides, once they choose side lengths, their creations are flexible. (This flexibility is sometimes called "play.") Triangles, on the other hand, have a kind of "rigidity" that is absent in polygons with more than three sides: one cannot make different triangles from a given set of side lengths.

Review the term **congruent** for polygons that are the same shape and size. In other words, two polygons are congruent if they are similar and if the ratio of corresponding sides is equal to 1.

Then ask students, Can you state what you learned in the activity, using the idea of congruence? If necessary, ask what would be true of the triangles formed if one student used the same three lengths as another student. With help, they should be able to formulate a statement like this:

If the sides of one triangle have the same lengths as the corresponding sides of another triangle, then the triangles must be congruent.

You may want to post this statement. *This principle is also expressed by saying that the set of side lengths "determines a triangle."* (You might recognize this as the SSS, or side-side-side, criteria for determining whether two triangles are congruent.)

Bring out that this statement does not hold for polygons in general. That is, if the sides of one polygon have the same lengths as the corresponding sides of another polygon, the two polygons are not necessarily congruent.

Although the activity uses the term *similarity,* the conclusions just stated are actually about congruence. For instance, in Question 3, students saw that if two triangles have identical side lengths, they must be congruent and therefore similar, with a ratio of corresponding sides that is equal to 1.

Ask students, What if the side lengths for two triangles are proportional, but not necessarily equal? They will probably conclude that the two figures will be similar. Assure them that this is true. You might post this conclusion as well:

If the corresponding sides of two triangles are proportional, then the triangles must be similar.

Then ask, Is it true that equal lengths of corresponding sides implies similarity for quadrilaterals? Students might imagine making one quadrilateral with a given set of lengths and another with sides all twice as long. They will probably see that the two quadrilaterals need not be similar, because the angles of either one could be changed without changing the side lengths. Thus, the following statement about polygons in general is true:

If the corresponding sides of two polygons are proportional, then the polygons are not necessarily similar.

Note that in the special case of triangles, two triangles are similar if and only if their corresponding sides are proportional. That is, similarity implies that corresponding sides are proportional, and corresponding sides being proportional implies similarity.

Key Questions

Can you state what you learned in the activity, using the idea of congruence?

What if the side lengths for two triangles are proportional, but not necessarily equal?

Is it true that equal lengths of corresponding sides implies similarity for quadrilaterals?

Supplemental Activities

Rigidity Can Be Good (extension) investigates the significance of geometric rigidity in the fields of architecture and construction.

Is It Sufficient? (reinforcement) is a geometric exploration and a continuation of ideas concerning logic and counterexamples.

Triangular Data (reinforcement) is another exploration about conditions for similarity, this time asking students to consider specific data for possible triangles.

More Similar Triangles

Intent

Students gain additional experience deriving and solving proportions for similar triangles. In this case, the triangles are overlapping.

Mathematics

In the lamp shadow model, there are two overlapping, similar triangles. This part of the unit continues to develop the skills students will need to recognize and relate the corresponding parts of these triangles. In this activity, students confront overlapping similar triangles for the first time. They learn to separate such similar triangles in order to find the unknown side lengths. In addition, work with more complicated, decimal proportions will show whether students understand the process of solving proportion equations.

Progression

Students address the tasks in this activity individually. First they "take apart" overlapping triangles to write equations and then solve them. Then they work with larger, decimal coefficients.

Approximate Time

15 minutes for activity (at home or in class)

10 minutes for discussion

Classroom Organization

Individuals, followed by whole-class discussion

Doing the Activity

You may want to make sure students see the overlapping similar triangles in the example. Suggest that it may be helpful for them to redraw each pair of overlapping triangles as separate triangles before deriving the proportion.

Discussing and Debriefing the Activity

Have volunteers present their answers for the four questions.

For Questions 1 and 2, have someone demonstrate how to redraw the triangles.

You may want to ask presenters for Questions 3 and 4 to explain the process they used to solve the proportion and then to draw a pair of triangles that these lengths may have come from. The triangles will not be to scale, but seeing a physical representation may help some students make the connection from the numbers to the side lengths.

Are Angles Enough?

Intent

Having learned that the shape of a triangle is determined by its side lengths, students now investigate whether a triangle is also determined by its angle measures.

Mathematics

If two triangles have the same angle measures, they have the same shape; that is, the two triangles are similar. (Because the sum of the angles in a triangle is fixed, only two angle measures need to be the same for one to draw this conclusion.) If, in addition to having the same angle measures, two triangles have sides connecting corresponding angles that are of the same length, then both the size and the shape of the triangles are the same.

Progression

Working in groups, students begin by drawing triangles with given angle measures and then compare their results. They repeat this for angle measures of their own choosing. Finally, they draw triangles with given angle measures and one given side length. They then talk about their work in a class discussion.

Approximate Time

40 minutes

Classroom Organization

Groups, followed by whole-class discussion

Doing the Activity

Students have seen, at least experimentally, that having corresponding sides proportional guarantees similarity for triangles. Now have them read and then do the activity in their groups.

Discussing and Debriefing the Activity

For Question 1, you might have some students trace their triangles onto transparencies. If so, choose examples of different sizes. Orient the transparencies so that the angles match up, superimposing the triangles. The nesting triangles will offer a vivid, visual confirmation of their similarity.

Students should recognize that there is a "rigidity" here, as they saw in *Why Are Triangles Special?* Although they can vary the size of the triangles in Question 1, the shape is determined.

In Question 3 students will see that once they choose one of the side lengths, they have no choice for the other two. (You might recognize this as the familiar ASA, or angle-side-angle, condition for triangle congruence.)

Post the general principle addressed in the activity:

If two triangles have their corresponding angles equal, then the triangles must be similar.

Some students may realize that only two pairs of corresponding angles need to be equal for similarity to exist, because the third angles are thus determined. If this comes up, ask students to explain their reasoning.

In Proportion

Intent

This activity gives students an opportunity to work with the concept of proportionality in nongeometric, real-world contexts.

Mathematics

Two similar triangles have side lengths that are proportional and angles that are equal. In other words, one similar triangle is related to another by a scale factor, but their angle measures remain invariant. Analogously, when proportions occur in the context of real life, multiplying one aspect of the situation by some factor doesn't mean everything should be multiplied by that factor. For example, when doubling a recipe, you double the amount of each ingredient but not the cooking time. In this activity, students continue to develop their understanding of similarity by focusing on the idea of invariance.

Progression

Students explore the three contexts posed here individually and discuss their results in class.

Approximate Time

15 minutes for activity (at home or in class)

20 minutes for discussion

Classroom Organization

Individuals, followed by whole-class discussion

Doing the Activity

This activity requires little or no introduction.

Discussing and Debriefing the Activity

The key idea to bring out in the discussion of this activity is that when one number in a situation is multiplied by some factor, other numeric aspects of the situation may or may not be multiplied by that factor.

After reviewing each example, ask how the ideas in this activity are related to what students have been learning about similar triangles. **How is this idea related to similar triangles?** Students should recognize that although, for example, the lengths of the sides of one triangle may be double those of a similar triangle, the angles are not doubled.

Key Question

How is this idea related to similar triangles?

Supplemental Activity

What If They Kept Running? (extension) uses distance and rate of speed as another context for investigating proportionality.

What's Possible?

Intent

Following investigations into why triangles are special and whether angles are enough to determine similarity, students now look at side lengths. They explore possibilities for side lengths for triangles and then try to generalize their discoveries to other polygons.

Mathematics

In any triangle, the lengths of any two sides must be longer than the length of the third side. That is, if a triangle has sides of lengths a, b, and c, then $a + b > c$, $a + c > b$, and $b + c > a$. This principle is known as the **triangle inequality**. (In the case of equality, the resulting triangles are "flat." In this activity, the triangle inequality is stated as a strict inequality.) This activity brings out the triangle inequality through an investigation and then generalizes the concept to other polygons.

Progression

Students work in groups to examine whether there are combinations of three side lengths that will not form a triangle. Then they investigate polygons with more than three sides. Groups share their findings in a class discussion.

Approximate Time

30 minutes

Classroom Organization

Groups, followed by whole-class discussion

Materials

Scissors

Straws, dry spaghetti, or strips of paper

Doing the Activity

To introduce the activity, point out that triangles have six "parts": three sides and three angles. You might ask whether the three angles can be any values. Students should recognize that the fact that the sum of the angles must be 180° limits the options.

Clarify that the goal of this activity is to examine whether the three sides of a triangle can be arbitrary lengths or whether there is some restriction on the possible combinations, perhaps analogous to the restriction on the angles.

As groups experiment and come up with the triangle inequality, have them move on to consider polygons with more than three sides.

Discussing and Debriefing the Activity

You might have one or two students present their groups' conclusions about triangles. **Can the sides of a triangle have any values? Why or why not?** Students should reach a conclusion equivalent to the following:

The sum of the lengths of any two sides of a triangle must be more than the length of the third side.

Tell students that this principle is called the **triangle inequality**. They may benefit from seeing this condition expressed symbolically, along with an appropriately labeled diagram.

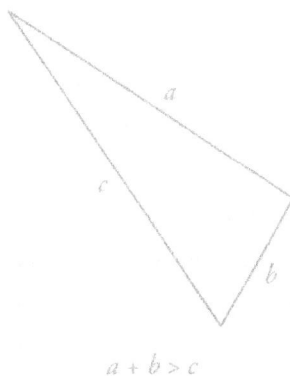

$$a + b > c$$

The triangle inequality: $a + b > c$ where a, b, and c can represent any sides of the triangle.

Someone may mention that to check whether three lengths are possible sides for a triangle, one only needs to verify that the two shorter lengths add to more than the longest one. If not, you might point this out yourself.

Then ask, **Did you find a similar generalization for quadrilaterals or other polygons?**

Students may have discovered a principle similar to the triangle inequality: that the longest side of a polygon must be less than the sum of the other sides.

Key Questions

Can the sides of a triangle have any values? Why or why not?

Did you find a similar generalization for quadrilaterals or other polygons?

Very Special Triangles

Intent

The primary purpose of this activity is to introduce right triangles and the basic terminology and concepts associated with them. Students will use this information later in the unit, when trigonometric functions are introduced.

Mathematics

Right triangles are the basis for trigonometry, and they appear in a wide range of geometry and measurement situations. In this activity, the basic parts of right triangles—the **hypotenuse**, the **legs**, and the angles adjacent to and opposite them—are introduced. Additionally, students examine the complementary relationship between the nonright angles, how the measures of angles relate to the lengths of the opposite sides, and some of the special conditions necessary for two right triangles to be similar.

Progression

Students explore the activity individually. The class will review the findings and then explore perpendicularity and obtuse and acute triangles.

Approximate Time

15 minutes for activity (at home or in class)

20 minutes for discussion

Classroom Organization

Individuals, followed by whole-class discussion

Doing the Activity

This activity requires little or no introduction.

Discussing and Debriefing the Activity

Review students' results. For Question 1, students may have a variety of explanations about why the other angles must be acute. Use this opportunity to review the principle, first developed in *Patterns* and encountered earlier in this unit (and proved later in this unit) that the sum of the angles of any triangle seems to be 180°.

This is also a good opportunity to bring out that the nonright angles of a right triangle must add to 90°. Introduce the term **complementary angles** for a pair of angles with a sum of 90°.

For Question 2, students will presumably see that the hypotenuse is the longest of the three sides. The **triangle inequality** was the focus of the previous activity, *What's Possible?,* so they may mention that as well. It is not the intent of Question 2 to bring out the Pythagorean theorem. That important result is discussed in Year 2 unit *Do Bees Build It Best?*

However, you may want to talk about the idea that the length of the hypotenuse is determined by the lengths of the two legs. Ask, Does knowing the lengths of the legs of a right triangle tell you anything about the length of the hypotenuse? Though students may not know the nature of the relationship is, they may recognize that there should be some function of the form $H = f(L_1, L_2)$, where H is the length of the hypotenuse and L_1 and L_2 are the lengths of the legs. (This fact is a consequence of the SAS, or side-angle-side, principle of congruence— that a triangle is determined by the lengths of two sides and the angle—the "included angle"—formed by those sides.)

Presumably students will see in Question 3 that the ratio of hypotenuse lengths is also 2, that the corresponding angles are equal, and that the two triangles fit the conditions for similarity. You can summarize this into another general principle of similarity:

If the legs of two right triangles are proportional, then the triangles are similar.

Question 4 is mainly an opportunity for using the terms **opposite side** and **adjacent side.** Students will probably have no problem with the principle that the longer leg is opposite the larger angle. But it should be interesting to see what sort of explanations they offer for this phenomenon. (This idea will become more relevant when students encounter the law of cosines in Year 3.) You might ask, Is the longest side opposite the largest angle in *every* triangle?

Question 5 is another look at the issue of angle sums. Students should be able to figure out that a 45°-45°-90° triangle is both isosceles and right and that there cannot be an equilateral right triangle.

The discussion of right triangles is a good time to bring up the term **perpendicular**. You might simply ask what two lines that meet at a right angle are called.

Introduce the symbol $\perp$ for perpendicularity. For example, tell students that one can express the fact that $\overline{AC}$ and $\overline{BC}$ in the activity are perpendicular by writing $\overline{AC} \perp \overline{BC}$.

As needed, review the terms **acute angle** and **obtuse angle**. Explain that a triangle with an obtuse angle is called an **obtuse triangle** and that a triangle whose angles are all acute is called an **acute triangle**.

Key Questions

Does knowing the lengths of the legs of a right triangle tell you anything about the length of the hypotenuse?

Is the longest side opposite the largest angle in *every* triangle?

What other kinds of angles are there besides right angles?

Angle Observations

Intent

Students explore via measurement some important relationships among angles formed by intersecting lines.

Mathematics

In the discussion of *Very Special Triangles,* students were introduced to **complementary angles**. Now they will explore the relationships among **supplementary angles**, **straight angles**, and **vertical angles**. Through their measurements, students might conjecture that vertical angles are equal in measure. The discussion of the activity will confirm that because vertical angles are supplements of the same angle, they must be equal.

Progression

In their groups, students measure angles in the given diagram and make conjectures about which angles are equal. They also investigate angle sum relationships and then generalize their results. A class discussion then introduces the terms *supplementary angles, straight angles,* and *vertical angles*.

Approximate Time

30 minutes

Classroom Organization

Groups, followed by whole-class discussion

Doing the Activity

The discussion that follows assumes that students are doing the activity by hand. If they have access to dynamic geometry software, however, they may want to work on this activity using the software rather than drawing diagrams on paper and measuring angles with a protractor.

You may want to spend a few minutes discussing angle notation. Point out that the standard practice is to identify an angle by naming three points in sequence, with the vertex as the middle point and one point from each ray forming the angle as the other two points.

In some cases, angles can be described unambiguously by a single letter. For example, in the diagram from the activity, reproduced below, writing $\angle A$ is clear because only one angle has A as its vertex. On the other hand, writing $\angle C$ is ambiguous, because it could refer to any of several angles. (We will use notation such as $\angle C$, rather than $m(\angle C)$, the "measure" of angle C, to represent the size of an angle.) In this case, the angles must be named using three letters, such as $\angle ACD$ or $\angle DCG$.

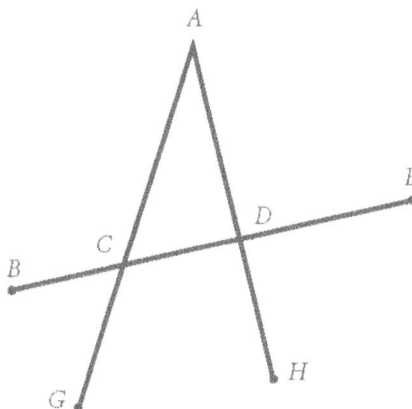

Warn students to draw the line segments long enough to allow them to measure the angles accurately.

Discussing and Debriefing the Activity

Ask several students to share their observations and explanations. As examples arise, introduce the appropriate terminology. For example, someone is likely to point out that $\angle ACD = \angle BCG$. If not, ask how each of these angles is related to $\angle BCA$. **How are angles *ACD* and *BCG* related to angle *BCA*?** If needed, prompt students by then asking, **If you knew that angle *BCA* was 110°, what other angles could you find?** Students should be able to state these relationships:

$$\angle ACD + \angle BCA = 180° \qquad \angle BCG + \angle BCA = 180°$$

Explain that $\angle ACD$ and $\angle BCA$ are called **supplementary angles** and that each angle is the *supplement* of the other.

Also introduce the term **straight angle** for an angle of 180°—that is, an angle whose two sides go in opposite directions. Point out that a pair of angles that fit together to form a straight angle are supplementary.

Introduce the term **vertical angles** for a pair of "opposite" angles formed by the intersection of two lines. Ask if anyone can state a general principle about such angles. Students should be able to see that this statement holds true:

Vertical angles are equal.

Taken together, the two equations above prove that $\angle ACD$ and $\angle BCG$ are equal. You may want to introduce the following observation, which is essentially the proof of the statement above.

Angles that are supplements of the same angle are equal.

Key Questions

How are angles *ACD* and *BCG* related to angle *BCA*?

If you knew that angle *BCA* was 110°, what other angles could you find?

More About Angles

Intent

Students investigate the relationships among angles formed by a transversal through two parallel lines.

Mathematics

When a transversal cuts two parallel lines, it creates a collection of angles in the region interior to, or between, the parallel lines, and in the regions exterior to these lines. **Corresponding angles** in the resulting diagram are equal, as are **alternate interior angles**.

Progression

Students work on the activity individually and share their results in class.

Approximate Time

15 minutes for activity (at home or in class)

15 minutes for discussion

Classroom Organization

Individuals, followed by whole-class discussion

Doing the Activity

This activity requires little or no introduction.

Discussing and Debriefing the Activity

Have several students share their observations about the angles in the diagram. They should note, for example, that $\angle ACD$ and $\angle CFG$ are equal and that $\angle BCF$ and $\angle CFG$ are equal.

Tell students that there is standard terminology for referring to certain pairs of angles in a diagram like this one.

You might introduce this terminology by identifying a pair of angles, such as $\angle ACD$ and $\angle CFG$, as corresponding angles and asking, What other pairs of angles in this diagram do you think are corresponding angles?

Using a diagram can be helpful for explaining why these angles are called corresponding. For example, $\angle ACD$ and $\angle CFG$ (labeled 1 and 2 below for clarity) are each "above and to the right" of their vertices, C and F.

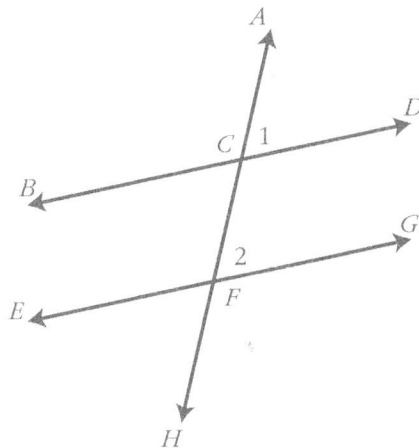

You might then introduce the term **alternate interior angles**, perhaps using $\angle BCF$ and $\angle CFG$. You can refer to the area between the parallel lines as the interior of the diagram and point out that $\angle BCF$ and $\angle CFG$ (angles 3 and 2 below) are in this interior region and on "alternate" sides of the transversal.

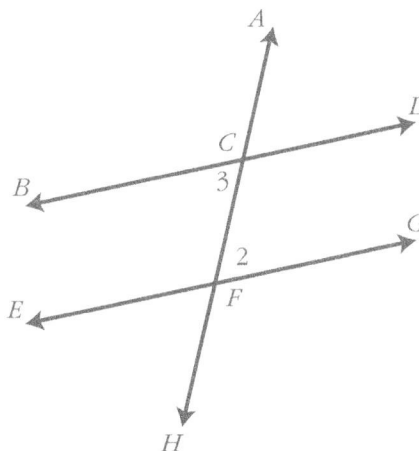

Try to get students to develop general statements like these:

When a transversal cuts across parallel lines, corresponding angles are equal.

When a transversal cuts across parallel lines, alternate interior angles are equal.

At some point, ask, How do you know that corresponding angles or alternate interior angles are equal? One likely response is that students measured the angles and the measurements were the same.

Key Questions

What other pairs of angles in this diagram do you think are corresponding angles?

How do you know that corresponding angles or alternate interior angles are equal?

Supplemental Activity

An Inside Proof (extension) asks students to prove two statements about a triangle and a line segment drawn through the triangle parallel to one of the triangle's sides.

POW 15: Trying Triangles

Intent

The third POW of the unit combines the triangle inequality with the basic ideas of probability developed in *The Game of Pig.*

Mathematics

In *The Game of Pig,* probability is defined as

$$P(\text{outcome}) = \frac{\text{the number of outcomes you're interested in}}{\text{the total number of possible outcomes}}$$

To find the probability of a particular outcome, you would count the number of outcomes you are interested in and the total number of possible outcomes, and then compute this ratio.

In this activity, both of these numbers are infinite, as the randomly point X could land in an infinite number of positions along the line segment. Students will have to use their understanding of the **triangle inequality** to find the portion of the segment that will produce a triangle and compare that length to the length of the entire segment. Thus, students will be finding a ratio of lengths.

Progression

Students work on the activity individually and share their results in class.

Approximate Time

15 minutes for introduction

1 to 3 hours for activity (at home)

15 minutes for presentations

Classroom Organization

Individuals

Materials

Pipe cleaners or something similar

Doing the Activity

You may want to demonstrate how to make a triangle from three segments of a pipe cleaner.

Discussing and Debriefing the Activity

Ask three students to make their presentations. If other students have different answers or explanations, let them share their ideas.

The key idea here is the use of the **triangle inequality**, which says that the length of the longest side of a triangle must be less than the sum of the lengths of the other two sides. Thus, students should see that both BX and XC must be more than $\frac{1}{4}$ of BC. (If BX is less than $\frac{1}{4}$ of BC, then $AB + BX < XC$; if BX is more than $\frac{3}{4}$ of BC, then $AB + XC < BX$.) This means that X must fall in the "center half" of $\overline{BC}$ for the three segments to be able to form a triangle.

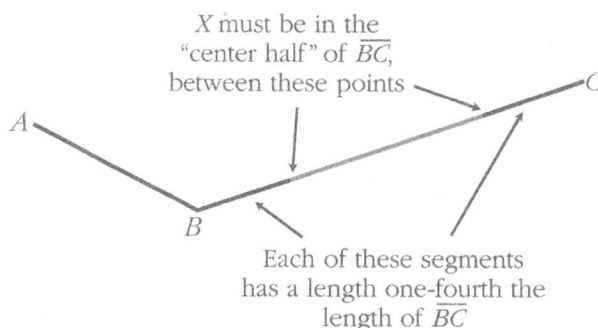

X must be in the "center half" of $\overline{BC}$, between these points

A

C

B

Each of these segments has a length one-fourth the length of $\overline{BC}$

The portion of $\overline{BC}$ along which point X may fall to produce a triangle is half of BC, so the probability that a triangle can be made is $\frac{1}{2}$.

If students raise the question of what happens if point X is exactly at the $\frac{1}{4}$ or $\frac{3}{4}$ point along $\overline{BC}$, you might mention the fact that when there are infinitely many equally likely results, the probability of any single result is zero. This statement will likely raise as many questions as it answers, but it may lead to an interesting discussion.

Inside Similarity

Intent

As this unit prepares to revisit the lamp shadow model, students use their results from *More Similar Triangles* and *More About Angles* to investigate how to form smaller, similar triangles within a larger triangle by drawing lines parallel to one of the larger triangle's sides.

Mathematics

If one draws a line segment connecting two sides of a triangle such that the segment is parallel to the third side, the resulting smaller, overlapping triangle is similar to the larger one. Students use the relationships among corresponding angles to establish this fact. This investigation will contribute to students' understanding of the relationships in the lamp shadow model, which consists of two overlapping, similar right triangles.

Progression

Students complete the activity individually, share findings in their groups, and then discuss the results as a class.

Approximate Time

5 minutes for introduction

15 minutes for activity (at home or in class)

20 minutes for discussion

Classroom Organization

Individuals, then groups, followed by whole-class discussion

Doing the Activity

Go over the activity as a class to make sure students understand what is being asked of them.

Discussing and Debriefing the Activity

Ask students to work in their groups to compile a list of ways to form a similar, small triangle inside a larger triangle. Then ask groups to report their findings.

They should have noted that a line through a triangle parallel to any side will produce a similar, small triangle. The diagrams illustrate the three families of lines that fit this description.

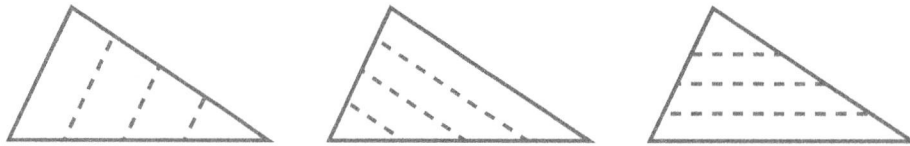

Ask students, **Why does a line parallel to one side of a triangle form a similar, smaller triangle?** They should be able to articulate that because the large and small triangles share one angle, and because the corresponding angles formed by the parallel lines are equal, the triangles have equal angles and are thus similar.

Key Question

Why does a line parallel to one side of a triangle form a similar, smaller triangle?

Supplemental Activities

Fit Them Together (reinforcement) prompts students to begin thinking about what happens to the area of a polygon when its dimensions are doubled.

Similar Areas (reinforcement) is a natural follow-up to the supplemental activity *Fit Them Together.*

A Parallel Proof

Intent

Building on their work in *More About Angles,* students are ready to construct one of the traditional proofs of the angle sum property for triangles.

Mathematics

The sum of the angles in a triangle is 180°. (This fact relies on the parallel postulate for its truth.) The proof developed here, which uses students' conjectures about the relationship between alternate interior angles formed by parallel lines cut by a transversal, is closely related to the investigation students did in *Degree Discovery*.

Progression

Students have been assuming that the sum of the angles in a triangle is 180°. They now work in groups to establish that this relationship is always true.

Approximate Time

25 minutes

Classroom Organization

Groups, followed by whole-class discussion

Doing the Activity

Have students work in groups on this activity. If a hint seems needed, you might ask, **Which angles must be equal? Which angle sum is easy to find?**

Discussing and Debriefing the Activity

After most groups seem to have found the proof, have one or two present their reasoning.

There are, basically, three steps to the argument.

- $\angle x = \angle s$ and $\angle y = \angle t$, because each is a pair of alternate interior angles formed by a transversal across parallel lines.

- $\angle x + \angle r + \angle y = 180°$, because these angles form a straight angle.

- Substituting $\angle s$ for $\angle x$ and $\angle t$ for $\angle y$ gives $\angle s + \angle r + \angle t = 180°$, as desired.

Key Questions

Which angles must be equal?

Which angle sum is easy to find?

Supplemental Activities

The Parallel Postulate (extension) offers students an opportunity to learn more about the history of the parallel postulate.

Exterior Angles and Polygon Angle Sums (extension) is an alternative proof of the angle sum property for triangles to the one in *A Parallel Proof,* based on the use of exterior angles.

Angles, Angles, Angles

Intent

In this final activity of *Triangles Galore,* students put to use their understanding of measurements of angles in polygons, in intersecting lines, and in parallel lines cut by a transversal.

Mathematics

Students will apply their knowledge of vertical angles, angles in triangles (including right triangles), and angles formed by a transversal across parallel lines to write and then solve equations to find missing angle measurements.

Progression

Students work on the activity individually and share their results in their groups and with the class.

Approximate Time

15 minutes for activity (at home or in class)

10 minutes for discussion

Classroom Organization

Individuals, then groups and whole-class discussion

Doing the Activity

Tell students that they now have the knowledge to solve these problems, which review the ideas they have been working with over the last few days.

Discussing and Debriefing the Activity

Have students check their answers in their groups. You may want to assign each group a question and ask them to share with the class the angle relationships in the diagram.

The Lamp Shadow

Intent

Equipped with an understanding of similarity, and specifically of similar triangles, students will now solve the first part of the unit problem: finding the length of a lamp shadow.

Mathematics

Similar polygons have equal corresponding angles and proportional corresponding sides. Either of these conditions ensures that two triangles are similar. Having used these ideas to set up and solve a variety of proportions, students now return to the lamp shadow model, which contains two overlapping, similar right triangles. These similar triangles allow the direct measurement of heights or distances that are difficult or impossible to measure directly.

By the end of *The Lamp Shadow,* students will have developed and used a function that relates shadow length, *S,* to the other three variables in the lamp shadow model.

Progression

The Lamp Shadow begins with a set of activities in which students use mirrors and lines of sight to create similar triangles that can be used to take indirect measurements. Several activities then build a solution to the lamp shadow problem. In addition, students present their results for the third POW and begin work on the fourth and last POW of the unit.

Bouncing Light

Now You See It, Now You Don't

Mirror Magic

Mirror Madness

A Shadow of a Doubt

To Measure a Tree

POW 16: Spiralaterals

More Triangles for Shadows

Bouncing Light

Intent

In this first of four activities that develop a technique for indirect measurement using mirrors and similar triangles, students explore what happens when light bounces off a mirror.

Mathematics

Students will discover through their experiments that the angle of approach of a light beam to a mirror equals its angle of departure. They will relate this principle to the use of a mirror to view the reflected image of an object.

Progression

Students conduct this in-class investigation in small groups, in which each student has a role, and share discoveries in a class discussion.

Approximate Time

30 minutes

Classroom Organization

Groups of 3

Materials

Flashlights (1 per group)

Mirrors (1 per group)

Doing the Activity

You may need to partially darken the room for this activity. It will also help if groups tape over the flashlight lens to produce only a sliver of light. This will make the beam less dispersed and easier to see as a "line" bouncing off the mirror.

Because a flashlight beam is more easily seen against a plain background, you may want to suggest that groups tape chart paper to their desks. They can then trace the light's path directly onto the paper.

The terms *angle of incidence* and *angle of reflection* refer to the angles between the light ray and a line perpendicular to the mirror, rather than between the light and the mirror itself. Because the angles between the ray of light and the mirror are more natural to work with, the student book speaks of the *angle of approach* and *the angle of departure*. Review the meanings of these terms with the class.

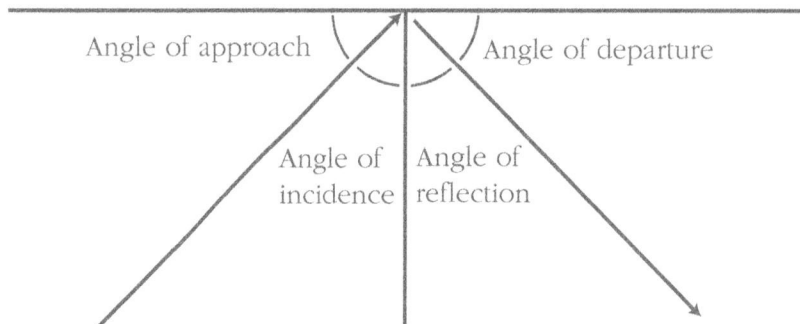

Angle of approach   Angle of departure

Angle of | Angle of
incidence | reflection

Students may have trouble knowing where to look in a mirror to find something that's out of sight, and they may be confused at seeing directions reversed. You may want to give groups some introductory tasks to help students develop an intuitive sense of how mirrors work. For example, place a mirror on the ground and ask students to stand so that they can see the top of the board in the mirror. Or ask one student to hold a mirror so that another can see an object—like a window, door, or poster—reflected in it. Have them explain how they are deciding where to place the mirror and what happens as they move it.

As groups make their drawings, suggest that they record the mirror's position as well as the path the light takes.

Discussing and Debriefing the Activity

When groups have had sufficient time to work with the flashlights and mirrors, bring the class together to discuss what they have learned. Theoretically, the angle of approach and the angle of departure should be equal. You might want to post the following observation:

Principle of light reflection: When light is reflected off a surface, the angle of approach is equal to the angle of departure.

Now You See It, Now You Don't

Intent
Students will use the principle that the angle of approach equals the angle of departure to analyze the line of sight to an object through a mirror.

Mathematics
When viewing the reflection of an object in a mirror, the line of sight follows the principle that the angle of approach is equal to the angle of departure. In this activity, students will locate these lines of sight and note that similar triangles are created.

Progression
Students investigate the two situations in this activity individually and share their results in class.

Approximate Time
15 minutes for activity (at home or in class)

15 minutes for discussion

Classroom Organization
Individuals, followed by whole-class discussion

Doing the Activity
Have students read the first question, and ask what they initially think might be the answer. Many will assume that only one letter can be seen.

Discussing and Debriefing the Activity
Discuss students' findings. Question 1 should be fairly straightforward, with students connecting point A to each end of the mirror and drawing the reflection lines.

For Question 2, ask, How did you get your point in Question 2? Watch for students who assume that the halfway point must be the point of reflection.

Which similar triangles are involved? How do you know the triangles are similar?

Students should be able to identify the similar triangles *RST* and *VUT* in a diagram like the one below (with spiders at points *R* and *V*) and prove their similarity by virtue of two equal angles: the pair of right angles as well as $\angle RTS$ and $\angle VTU$, which are equal by the principle of light reflection.

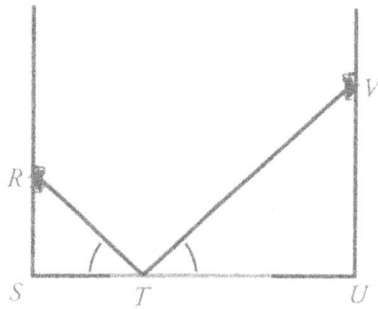

What fraction of the way from point *S* to point *U* is point *T*? Try to elicit the explanation that, because of the similar triangles, the lengths *ST* and *TU* must be in the same ratio as *RS* and *VU*.

Key Questions

How did you get your point in Question 2?

Which similar triangles are involved? How do you know the triangles are similar?

What fraction of the way from point *S* to point *U* is point *T*?

Mirror Magic

Intent

Students will incorporate ideas from the previous two activities to develop an indirect measurement method using mirrors and similar triangles.

Mathematics

When the height of an object is difficult to measure directly, one can use the following indirect measurement method. Place a mirror on the ground and move to a position where you can see the object in the mirror. Two similar right triangles are created in this situation. Your height, your distance along the ground from the mirror, and the object's distance along the ground from the mirror can be measured and used to form and solve a proportion to find the height of the object.

Progression

Students work on this activity in pairs and share discoveries in a class discussion.

Approximate Time

40 minutes

Classroom Organization

Pairs

Materials

Mirrors

Doing the Activity

After reading the introductory material as a class, you might have one or two students act out the scenario.

Assign classroom objects to pairs of students, or allow students to choose their own objects.

Students are being given less structure here than in earlier activities, so they may need a few hints to get started. In particular, they might need suggestions about how to use the principle of light reflection to create similar triangles. As much as possible, though, allow them to figure out on their own how to apply the concepts they have been learning to solve this problem.

It is hoped that, in creating their diagrams, students will find the appropriate similar triangles. In the setup diagrammed below, students can measure the person's height, the distance along the floor from the person to the mirror, and the distance along the floor from the mirror to the object, and then set up a proportion to find the object's height. The second diagram is a more schematic version of the first and may make the similar triangles more apparent.

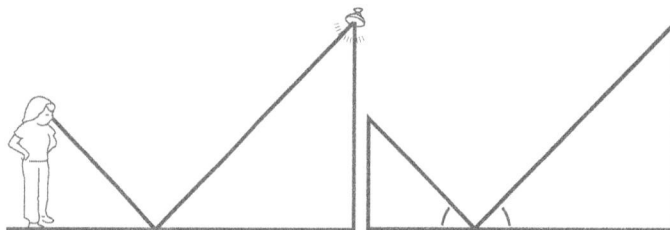

Discussing and Debriefing the Activity

Have students explain what information they need to calculate the height of their object and how they use this information. Keep the emphasis on the use of diagrams and on the mathematics of the problem, rather than on solving for the height of the object.

Students should be able to apply their knowledge of similar triangles to set up a proportion like the following:

$$\frac{\text{height of object}}{\text{height of person}} = \frac{\text{distance from mirror to object}}{\text{distance from mirror to person}}$$

The distances here are measured along the ground. (It would be mathematically correct to use the distances to the top of the object and the person, but these can't be measured easily.) Also note that height of the person is technically the height of the person's eyes.

Students should have used variables—on the diagrams and in the equations—to represent the four measurements. You might ask, **Which variables did you use for these measurements?**

Key Question

Which variables did you use for these measurements?

Mirror Madness

Intent

This playful activity, which concludes the first part of *The Lamp Shadow,* offers students more practice with similar triangles.

Mathematics

The questions in this activity can each be answered by using the reflection of lines of sight in a mirror to set up a pair of similar triangles and then write and solve a proportion.

Progression

Students work on the activity individually and share results in their groups and with the class.

Approximate Time

15 minutes for activity (at home or in class)

15 minutes for discussion

Classroom Organization

Individuals, then groups, followed by whole-class discussion

Doing the Activity

This activity requires little or no introduction.

Discussing and Debriefing the Activity

Have students share their findings in their groups. The amount of discussion needed for this activity will depend on how comfortable students were with the activity *Mirror Magic*. Again, focus on the use of similarity to set up the equations, rather than on equation-solving techniques.

To find the height of Momma spider, for example, students might create a diagram like this, with Sister at point *A* and Momma at point *B*.

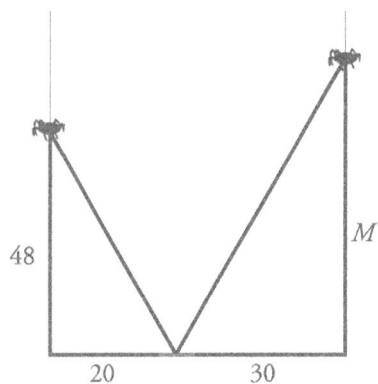

They should be able to explain why the two triangles are similar (using the principle of light reflection) and then use that similarity to set up an equation such as $\frac{20}{30} = \frac{48}{M}$. They may then use a variety of methods, including trial and error, to determine the value for M that fits the equation.

A Shadow of a Doubt

Intent

Students return to the lamp shadow problem and develop a general equation relating the variables in the model.

Mathematics

In *The Shadow Model,* students developed a mathematical model for the lamp shadow problem. In the activities in *The Shape of It,* they built an understanding of similar figures and, in *Triangles Galore,* they focused on the proportional relationships among corresponding sides of similar triangles. In *More Similar Triangles* and *Inside Similarity,* students encountered overlapping similar triangles, and in *Very Special Triangles,* they learned about right triangles. They will now bring all this work together to develop a general equation connecting the variables in the lamp shadow model. They also have their first opportunities to use this equation to find specific shadow lengths, *S,* for given values of *L, D,* and *H.*

Progression

Students work on the activity in groups, with class discussion of Questions 1 and 2 to support their work on Questions 3 to 5.

Approximate Time

25 minutes

Classroom Organization

Groups and whole class

Doing the Activity

Have students focus on the diagram that opens the activity. As needed, review the meaning of each variable.

- *L* is the distance from the light source to the ground.

- *D* is the distance along the ground from the light source to the object casting the shadow.

- *H* is the height of the object casting the shadow.

- *S* is the length of the shadow.

Mention that the emphasis in this activity is on the development and verification of an equation relating the four variables, and encourage the class to restate the unit goal for the lamp shadow model.

Unit goal: To find a formula expressing S, the length of a shadow, in terms of the variables L, D, and H.

Discussing and Debriefing the Activity

Groups must come up with the proportion in order to answer Questions 3 to 5, so you will probably want to bring the class together to discuss Questions 1 and 2.

You can begin by having a student summarize how to use the diagram to find a general equation relating the four variables. **How did you use the shadow diagram to get an equation relating *S*, *L*, *D*, and *H*? What similar triangles are involved?** Emphasize what the similar triangles are and how students can be sure that they are similar.

It is expected that students will see the similarity between the large and small triangles in the diagram below and write $\frac{S}{S+D} = \frac{H}{L}$ or an equivalent proportion (such as $\frac{L}{H} = \frac{S+D}{S}$). Some students may need help identifying the horizontal side of the large triangle as $S + D$.

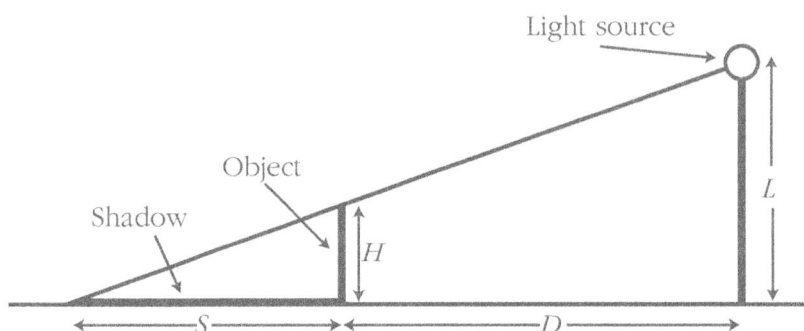

Once everyone has an equation, have groups continue work on the activity.

Have volunteers present solutions to Question 3 and 4. For Question 5, have a few students describe the process of finding the length of a shadow in their own words.

Key Questions

How did you use the shadow diagram to get an equation relating *S*, *L*, *D*, and *H*?

What similar triangles are involved?

To Measure a Tree

Intent

In this activity, students apply their knowledge of similar triangles in a real-world context.

Mathematics

Similar triangles allow the indirect measurement of heights or distances that are difficult or impossible to measure directly. In this activity, students develop methods, using pairs of similar triangles, to find the height of a tree. Using a line of sight from the top of the tree, to the top of a person, to the ground is one way to approach this problem; using the lengths of the shadows cast by the tree and by a person of known height is another.

Progression

Students work on the activity individually and then share methods in their groups and with the class.

Approximate Time

20 minutes for activity (at home or in class)

25 minutes for discussion

Classroom Organization

Individuals, then groups, followed by whole-class discussion

Doing the Activity

You may want to spend some time as a class brainstorming how students might approach the problem.

Discussing and Debriefing the Activity

You might ask students to take turns in their groups, explaining a method for measuring the height of a tree and then have each group choose one method they might present to the class. If logistics permit, you may want to take students outdoors to apply the methods to a tree or other tall structure.

There are several approaches to the problem, all based on similar triangles. Following are three possible methods and questions you might ask as the discussion proceeds.

Sample Method 1

This approach begins by mentally extending a line from the top of the tree, to the top of a person's head, to the ground.

If this method is presented, ask, **What are the triangles in this diagram? Are they similar? How do you know?** It will probably be helpful to draw a schematic diagram of the situation.

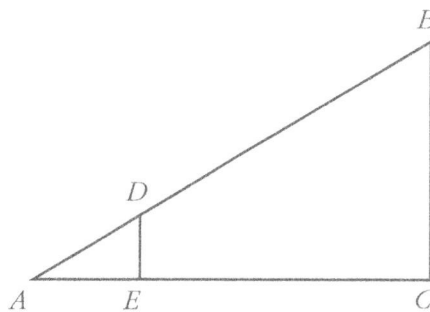

You might also draw the large and small triangles separately to identify them more clearly.

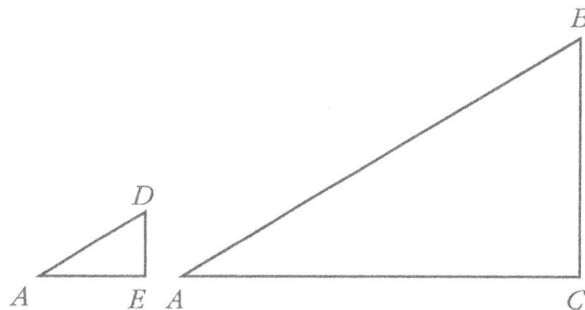

Students will probably demonstrate similarity by noting that the two triangles have $\angle A$ and a right angle in common.

Ask students to indicate the tree height and other lengths in the diagram and to identify which distances they would be able to measure more easily than the height of the tree. They should see that the tree height is BC and that they might be able to measure these distances:

- the height of the person (DE)

- the distance along the ground from the person to the tree (EC)

- the distance along the ground from the end of the diagonal to the tree (AC)

If not many groups were able to identify the similar triangles, you may want to let them work for a few minutes to develop a proportion, such as the following, that would allow them to find the tree height.

$$\frac{DE}{BC} = \frac{AE}{AC}$$

To find the length *AC*, one needs to first identify point *A*. One way a person could do this is to have a friend sight along the diagonal line that connects the top of the person's head and the top of the tree and mark where this line hits the ground.

Sample Methods 2 and 3

These two methods involve the use of shadows.

If the sun is shining, you can stand so that you are just barely in the tree's shadow and so that your shadow and the tree's shadow end at the same point. This way, the diagonal line will go to the end of the common shadow.

Another approach that involves shadows is to stand anywhere and measure your height, the length of your shadow, and the length of the tree's shadow. Then construct and solve the following proportion:

$$\frac{\text{your height}}{\text{tree's height}} = \frac{\text{your shadow}}{\text{tree's shadow}}$$

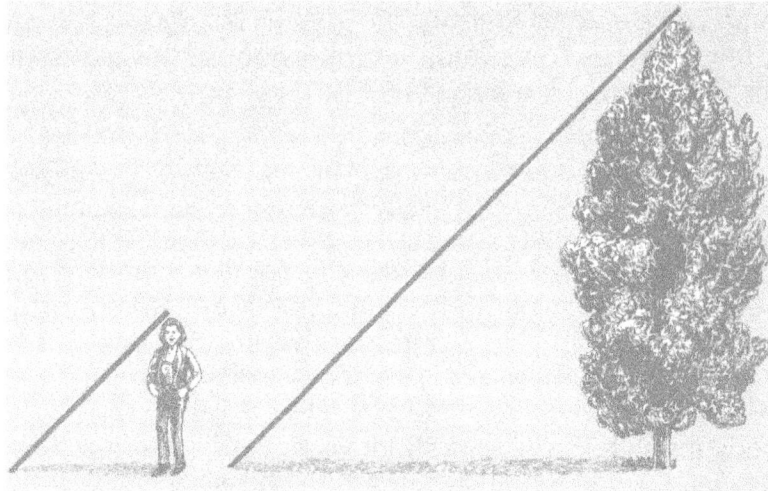

To check understanding, you might give the class specific values for the measurable distances. Have them focus on setting up a proportion, rather than on finding the solution. For example:

The woman measuring the tree is 5 feet tall and is standing 20 feet from the tree. Her friend finds that the line of sight from the top of the tree past the top of the woman's head hits the ground just 4 feet behind the woman. What is the height of the tree?

Key Questions

What are the triangles in this diagram? Are they similar? How do you know?

How might you use shadows to work on this problem?

POW 16: Spiralaterals

Intent

The last POW of the unit is a geometric investigation in which students look for patterns in the figures formed by line segments that reflect various sets of number sequences.

Mathematics

A "spiralateral" is the spiral-like shape formed by connected line segments generated by a sequence of numbers. In this open-ended investigation, students decide how and what to explore about these shapes and how they are related to the sequences that generate them. For example, they may decide to investigate the question of when and why a spiralateral returns to its starting point.

Progression

Students look for patterns and organize their information to make generalizations.

Approximate Time

15 minutes for introduction

1 to 3 hours for activity (at home)

20 minutes a week or so later for presentations

Classroom Organization

Individuals

Doing the Activity

Take some class time to illustrate how a spiralateral is made. Seeing an example will be much clearer for some students than reading a written description. Let students begin to explore the problem in their groups.

You may want to suggest areas of further exploration for interested students. For example:

- What happens if 0 is used in the sequence?

- Can negative numbers be used? What about fractions?

- How does the analysis change if the angle of turn is 60°? 120°?

Discussing and Debriefing the Activity

Ask three students to make POW presentations. As presenters share their observations about spiralaterals, focus the class's attention on explanations of the discoveries.

For example, if students investigated the question of when a spiralateral returns to its starting point, they may have seen that sequences of length 2, 3, or 5 always return, but sequences of length 4 do not necessarily return. If so, you might ask, **Why might sequences of length 4 be different from sequences of other lengths? Are other lengths for which spiralateral sequences do not always return to the start?**

After the initial presentations, ask other students to share any discoveries they made or any variations on spiralaterals that they investigated.

Key Question

Why might sequences of length 4 be different from sequences of other lengths?

Are other lengths for which spiralateral sequences do not always return to the start?

More Triangles for Shadows

Intent

Students will investigate the geometry of the lamp shadow situation in more depth in order to develop an equation for shadow length in terms of the other variables.

Mathematics

The relationship among the four variables in the lamp shadow model found by students in *A Shadow of a Doubt* does not express shadow length, S, as a function of the other three variables, H, D, and L. It is possible to solve the relationship for S to get this function, but students are just developing this manipulative facility. In addition, finding the function using algebraic manipulation will not explain *why* this relationship holds. To find this function from the geometry of the model, students will draw an auxiliary line to create a new pair of similar triangles.

Progression

Students work on this activity in groups and as a class.

Approximate Time

25 minutes

Classroom Organization

Groups and whole class

Doing the Activity

Have volunteers read the activity aloud.

Ask students to analyze the diagram in their groups. Mention that there is another triangle in the diagram that can be used to help find the equation.

To form that triangle, students will have to add another line to the diagram. Specifically, ask them to look for a triangle that has a side of length D.

Show students where to draw the new line if they are unable to discover it themselves.

Discussing and Debriefing the Activity

Have a volunteer describe how to get the equation. The key is to add a horizontal line to the diagram, of length D, and then to use the similarity of the two small triangles.

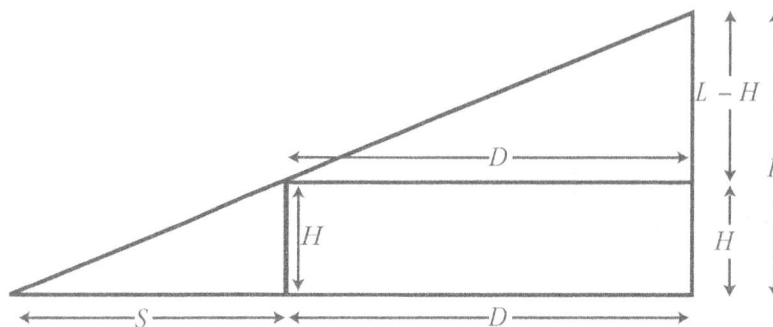

Ask students, **How do you know that these triangles are similar?** Their explanations will probably involve their work with **corresponding angles** in *More About Angles*.

Groups should then be able to form a proportion equation, like this one:

$$\frac{S}{D} = \frac{H}{L - H}$$

The final step is to solve for *S* and get something like

$$S = \frac{DH}{L - H} \quad \text{or} \quad S = D\frac{H}{L - H}$$

Key Questions

Is there another triangle you can find in this diagram that would help?

How do you know that these triangles are similar?

The Sun Shadow

Intent

The Sun Shadow wraps up the unit by developing the mathematics students need to solve the second part of the unit problem: finding the length of a sun shadow.

Mathematics

The length of a shadow created by the sun is a function of the height of the object casting the shadow and the angle of elevation of the sun. A mathematical model of this situation is a right triangle, with the object and the shadow as the legs and the angle of elevation as one of the acute angles. The length of the shadow can then be found using trigonometry.

These activities introduce the basics of trigonometry, anchored in students' understanding of similarity. Students then use these ideas to develop a function that relates shadow length, *S,* to the other variables in the sun shadow model.

Progression

Once *The Sun Shadow* reintroduces the sun shadow problem, a set of six activities develops the basic ideas of trigonometry. Then students use trigonometry to find the function that solves the second part of the unit problem. In addition, they present their results on the last POW, complete unit assessments, and compile a unit portfolio that also looks back over the entire year.

The Sun Shadow Problem

Right Triangle Ratios

Sin, Cos, and Tan Revealed

Homemade Trig Tables

Your Opposite Is My Adjacent

The Tree and the Pendulum

Sparky and the Dude

A Bright, Sunny Day

Beginning Portfolio Selection

Shadows Portfolio

The Sun Shadow Problem

Intent

Students return to the sun shadow problem with a goal of identifying the key variables in the situation. The sun shadow problem will actually be solved in the activity *A Bright, Sunny Day*.

Mathematics

A shadow cast by the sun differs from one cast by a lamp in an important way, which students observed early in this unit: the sun shadow doesn't change length as the object casting the shadow moves. This is because the source of the light is, in effect, infinitely far away. The length of a sun shadow is a function of just two variables: the height of the object casting the shadow and the **angle of elevation** of the sun.

Progression

Students discuss the sun shadow situation as a class, sharing ideas and working to identify the key variables in the model.

Approximate Time

25 minutes

Classroom Organization

Whole class

Doing the Activity

As a class, read the activity in the student book.

Discussing and Debriefing the Activity

Ask students to share their ideas on this situation. They should recognize that the height of the object casting the shadow, which has been labeled *H* previously, is still an important variable.

They might also mention time of day and position on the globe as variables that could affect the length of a sun shadow. Ask, **Why might "time of day" or "position on the globe" affect the length of a sun shadow?** Try to get students to see that these things affect the angle at which light from the sun hits an object. The goal is for them to recognize that the angle of the sun's position is the other crucial variable.

It may help to bring out that at noon, sun shadows are at their shortest, and that toward dusk or shortly after dawn, sun shadows are at their longest.

If students have trouble picturing the desired angle, suggest that they look straight ahead and then tilt their heads as if looking up toward the sun. The amount of "tilt" is the angle in question.

Culminate this discussion by introducing a diagram similar to the one below. Introduce the term **angle of elevation** for the "tilt" represented by θ (the lowercase Greek letter theta) in this diagram. Students should recognize that they are looking for a way to express S as a function of H and θ. That is, they want a formula for a function g such that $S = g(H, \theta)$.

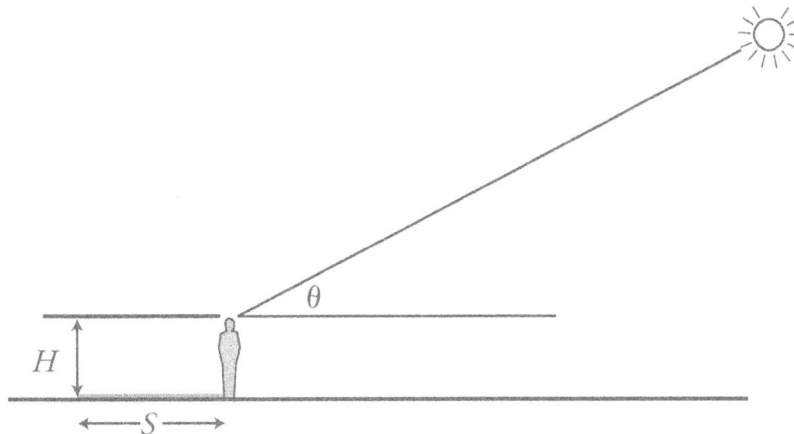

Key Question

Why might "time of day" or "position on the globe" affect the length of a sun shadow?

Right Triangle Ratios

Intent

Right Triangle Ratios begins a sequence of six activities in which students are introduced to trigonometric ratios through concepts of similarity and then apply those ratios to solve problems. In this activity, students create their own right triangles, find several ratios, and then compare those ratios with their classmates. This work provides the basis for the introduction of the three primary trigonometric functions for right triangles.

Mathematics

If two right triangles have an acute angle in common, then the triangles are similar and the ratios of corresponding sides are equal. In this activity, students will draw right triangles with an acute angle of 55°. Using this angle as a reference, they then calculate three ratios: $\dfrac{\text{opposite side}}{\text{hypotenuse}}$, $\dfrac{\text{adjacent side}}{\text{hypotenuse}}$, and $\dfrac{\text{opposite side}}{\text{adjacent side}}$.

Finally, they compare their results, which will be about the same no matter the exact size of the right triangle.

Progression

Students work on the activity individually, compare results in their groups, and then review the ideas in a class discussion.

Approximate Time

20 minutes for activity (at home or in class)

15 minutes for discussion

Classroom Organization

Individuals, then groups, followed by whole-class discussion

Doing the Activity

Introduce the activity, emphasizing that students need to measure the sides of their triangles as carefully as possible in order to get fairly accurate ratios in Question 2.

Discussing and Debriefing the Activity

Have students share the ratios they computed in their groups. Then bring the class together for a discussion of their discoveries.

Students should have found that even with triangles of very different sizes, the ratios come out essentially equal. In other words, they may get different values for Question 1 but approximately the same answers for Question 2. If this is not the case, you may need to review the accuracy of their measurements or the computation of the ratios.

Ask for an explanation. **Why did you all get approximately the same ratios?** Students should mention the principle of similarity and be able to explain why the triangles are all similar: they are all right triangles, as each has a right angle, and they all have a 55° angle.

For emphasis, ask, **Is there anything special about having an angle of 55°, or do our results illustrate a more general principle?** Help students come up with something like this statement:

If two right triangles have an acute angle in common, then the triangles are similar.

Key Questions

Why did you all get approximately the same ratios?

Is there anything special about having an angle of 55°, or do our results illustrate a more general principle?

Sin, Cos, and Tan Revealed

Intent

This reference page introduces students to the trigonometric functions of **sine**, **cosine**, and **tangent**. Students investigated these ratios in *Right Triangle Ratios*.

Mathematics

The three trigonometric functions of sine, cosine, and tangent are defined in this reference page as specific ratios of sides of a right triangle. Students' introduction to these terms is anchored by the concept of similarity.

Progression

Students review the presented information in a whole-class discussion.

Approximate Time

10 minutes

Classroom Organization

Whole class

Using the Reference Page

Tell students that the ratios within right triangles that they examined in *Right Triangle Ratios* are part of the branch of mathematics called **trigonometry** and that each of these ratios has a name. You may want to point out that these ratios are important in part because right triangles are so important.

Have volunteers read aloud sections of the reference page, periodically pausing to make sure students understand what is being said.

Homemade Trig Tables

Intent

Students investigate and build their own tables of values for the ratios of sine, cosine, and tangent.

Mathematics

Trigonometry is built on the concept of similarity as applied to right triangles. As students construct a class table of trigonometric values in this activity, they will be reminded that, because of similarity, for any acute angle θ, the ratio called sine, for example, is the same no matter how large the right triangle containing θ is drawn.

Progression

Working in pairs, students extend their experience of finding ratios in a right triangle with a 55° angle to finding ratios in a right triangle with angles from 10° to 80°. They compile their results in a class table.

Approximate Time

35 minutes

Classroom Organization

Pairs, followed by whole-class discussion

Doing the Activity

Inform students that a long time ago, people calculated the ratios of sine, cosine, and tangent for a wide variety of angles, recorded the ratios in tables, and then published the information. The information in those "trigonometric tables" is now available in calculators—more conveniently, more comprehensively, and more precisely.

You may want to remind students that the terms opposite side and **adjacent side**

are always used in relation to a specific angle. For example, in triangle *ABC*, *BC* is adjacent to $\angle B$ and opposite $\angle A$.

Have students work on the activity in pairs. Assign several angles to each pair, from 10° to 80° in multiples of 10°. If possible, assign each angle to at least two pairs.

Prepare a table on chart paper for students to fill in as they find the ratios.

Angle	Sine of the angle	Cosine of the angle	Tangent of the angle
10°			
20°			
30°			
40°			
50°			
60°			
70°			
80°			

Discussing and Debriefing the Activity

Allow time for students to examine the accumulated results and look for patterns in each column of the trig table. The visual display will offer insight into how the trigonometric functions behave. Ask for volunteers to share their observations.

What do you notice about the behavior of any of the trigonometric functions?

What's the smallest value that the sine of an angle can have? The largest? What about cosine? What about tangent?

As these facts are expressed, ask students to think about why they make sense. Help them to connect the behavior of each function directly to the geometry of the right triangle. For example, ask, **As the angle increases, what happens to the ratios of the sides? For example, as the acute angle goes from 10° to 80°, what does the sine ratio do?**

You might also want to ask whether students see any patterns by comparing one trigonometric function to another.

Key Questions

What do you notice about the behavior of any of the trigonometric functions?

What's the smallest value that the sine of an angle can have? The largest value? What about cosine? What about tangent?

Your Opposite Is My Adjacent

Intent
This activity follows the introduction of the trigonometric functions with an investigation into the relationship between sine and cosine.

Mathematics
The sine and cosine are cofunctions, short for complementary functions. These functions are defined at this point as ratios of sides in a right triangle. In any right triangle, the side opposite one acute angle is adjacent to the other acute angle, and these two angles are **complementary angles**. If $\angle A$ and $\angle B$ are the two acute angles, then the side opposite $\angle A$ is adjacent to $\angle B$, and sin A, which is the ratio $\dfrac{\text{angle opposite } \angle A}{\text{hypotenuse}}$, is equal to cos B, which is the ratio. Because $\angle A$ and $\angle B$ are complementary, $B = 90° - A$ and $\sin A = \cos(90° - A)$.

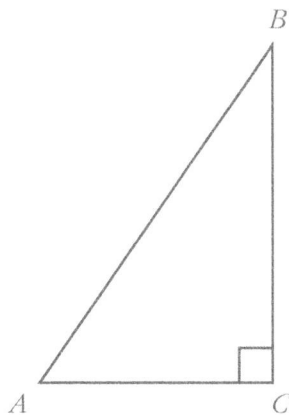

Progression
Students work on the activity individually and share results in their groups and with the class.

Approximate Time
15 minutes for activity (at home or in class)

15 minutes for discussion

Classroom Organization
Individuals, then groups, followed by whole-class discussion

Doing the Activity
This activity requires little or no introduction.

Discussing and Debriefing the Activity

Have students share results in their groups. You might ask each group to be ready to answer one of the questions.

For Question 1, students should see that $\angle A + \angle B = 90°$ or, equivalently, $\angle A = 90° - \angle B$ or $\angle B = 90° - \angle A$. For the discussion of Question 3, it will be helpful if they see this relationship stated in all three ways. You might bring out the equations $\angle A = 90° - \angle B$ and $\angle B = 90° - \angle A$ by asking, **How could you get one angle if you knew the other?**

Review the term **complementary angles** (first introduced in the discussion of *Very Special Triangles*) for a pair of angles with a sum of 90°.

For Question 2, students should recognize that this ratio is both sin *A* and cos *B*. You can take this opportunity to point out that the word *cosine* begins with "co," as does the word *complementary*.

For Question 3, students should be able to put Questions 1 and 2 together to get the general formulas

$$\sin A = \cos(90° - A) \qquad \cos A = \sin(90° - B)$$

Key Question

How could you get one angle if you knew the other?

The Tree and the Pendulum

Intent

In this activity and the next, students apply their new knowledge of trigonometry to solve "missing side" problems. Their work on these activities will offer an indication of how well they have absorbed the basics of trigonometry.

Mathematics

The trigonometry of right triangles allows one to compute the remaining sides and angle of a right triangle if the measures of one side and one acute angle are known.

Progression

Students work on the activity in groups and share results with the class.

Approximate Time

25 minutes

Classroom Organization

Groups, followed by whole-class discussion

Doing the Activity

Students' main difficulty in this activity will be deciding which trigonometric function to use. If so, refer them to the definitions of the ratios and clarify again the meanings of the terms opposite and adjacent.

Discussing and Debriefing the Activity

You may want to have volunteers present their work on the two questions.

For Question 1, students should be able to set up a diagram like the one below and see that the ratio $\frac{x}{12}$ is equal to tan 70°, so $x = 12(\tan 70°) \approx 32.97$. Adding the distance from Woody's eyes to the ground (5 feet) gives the height of the tree as approximately 38 feet.

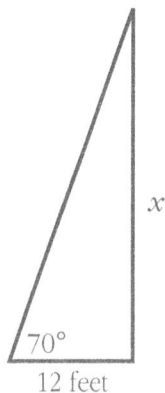

Question 2 involves the sine function. Students should be able to set up an equation like $\sin 30° = \dfrac{d}{30}$, which is equivalent to $d = 30(\sin 30°)$, giving a distance of 15 feet.

Supplemental Activities

Exactly One-Half! (extension) asks students to prove that sin 30° is exactly 0.5.

Eye Exam and *Lookout Point* (reinforcement) presents two real-world problems that can be solved using trigonometric functions.

Sparky and the Dude

Intent

This activity give students more experience using trigonometric ratios to find distances.

Mathematics

As in *The Tree and the Pendulum*, students will use the idea that the remaining sides and angle of a right triangle can be determined if one side length and one acute angle are known. In the first question, they will encounter an angle for a line of sight measured down from the horizontal, sometimes called the *angle of depression* to distinguish it from the *angle of elevation*. Applying what they learned in *More About Angles,* students will notice that the angle of depression from one vantage point is equal to the angle of elevation from the other.

Progression

Students work on the activity individually and share results in a class discussion.

Approximate Time

10 minutes for activity (at home or in class)

15 minutes for discussion

Classroom Organization

Individuals, followed by whole-class discussion

Doing the Activity

Students will need a calculator with trigonometric functions for this activity.

Discussing and Debriefing the Activity

For Question 1, ask the class, **What triangle did you use?** There are two identical triangles in the situation, labeled Triangle I and Triangle II in the diagram below.

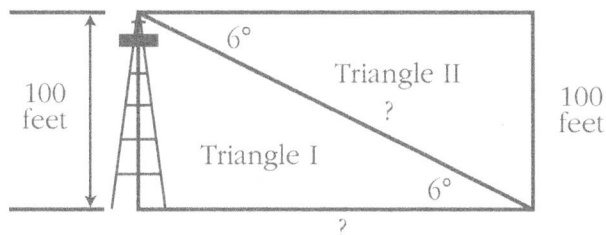

If students work with Triangle I, they need to recognize that the angle at the lower right is 6°. If they work with Triangle II, they need to realize that the length of the right side is 100 feet and that the length of the top side is the same as the distance from the base of the tower to the fire.

Based on one triangle or the other, students should have come up with these relationships:

$$\sin 6° = \frac{100 \text{ feet}}{\text{distance from Sparky to fire}}$$

$$\tan 6° = \frac{100 \text{ feet}}{\text{distance from base of tower to fire}}$$

If students question the effect of Sparky's own height, you might have them recalculate and then compare the values, using a height of 5 feet for Sparky. Then ask how significant Sparky's height is when sighting a fire 1000 feet away.

Students should have been able to find the values of sin 6° and tan 6° from their calculators, and then, with some trial and error, the desired distances. It's about 951.4 feet from the fire to the base of the tower and about 956.7 feet from the fire to Sparky.

Some students may have used the complementary angle, 84°, to compute the distance from the fire to the base of the tower by using the equation

$$\tan 84° = \frac{\text{distance to fire}}{100 \text{ feet}}$$

The distance from the fire to the base of the tower is thus 100(tan 84°), or again about 951.4 feet.

The basic equation for Question 2 is

$$\tan 28° = \frac{\text{height of cliff}}{50 \text{ meters}}$$

This gives the height of the cliff as approximately 26.6 meters. Since Dave's height and Charlene's height are equal, they cancel out.

Key Question

What triangle did you use?

Supplemental Activities

Pole Cat (reinforcement) reinforcement is another standard trigonometry problem.

Dog in a Ditch (reinforcement) looks at first glance like a trigonometry exercise. However, at this stage it's likely that students will have to solve this problem by making scale drawings. (The law of sines provides the simplest solution, but students don't know about it yet.)

A Bright, Sunny Day

Intent

The mathematical work of the unit concludes with this activity, in which students put their knowledge of trigonometry to use to find the solution to the second part of the unit problem: finding a formula for the length of a sun shadow.

Mathematics

The length of a sun shadow is a function of two variables: the height of the object casting the shadow and the angle of elevation of the sun. The object, the shadow, and the line of sight to the sun create a right triangle. The length of the shadow can be found using trigonometry.

Progression

Students work on this final activity in groups and share their results as a class.

Approximate Time

25 minutes

Classroom Organization

Groups, followed by whole-class discussion

Doing the Activity

You may want to review the diagram in the student book with the class. You might suggest that groups start with specific values for H and θ, which may help them figure out how to solve the trigonometric equation for S in terms of H and θ.

If necessary, tell students that the angle of elevation from their eyes to the sun is the same as the angle in the triangle formed by their bodies and their shadows (both labeled θ in the diagram below).

Discussing and Debriefing the Activity

There are essentially three steps to solving this problem. You may want to have presentations on each.

- Setting up a clear, correctly labeled diagram

- Using the diagram to get an equation involving S, H, and θ

- Solving this equation for S in terms of H and θ

Once students have a diagram like the one below, they will probably start with the

equation $\tan q = \dfrac{H}{S}$, which they must then solve for S. It may help to work this out once or twice with specific values for *H* and θ and then use those examples to develop the general equation $S = \dfrac{H}{\tan q}$.

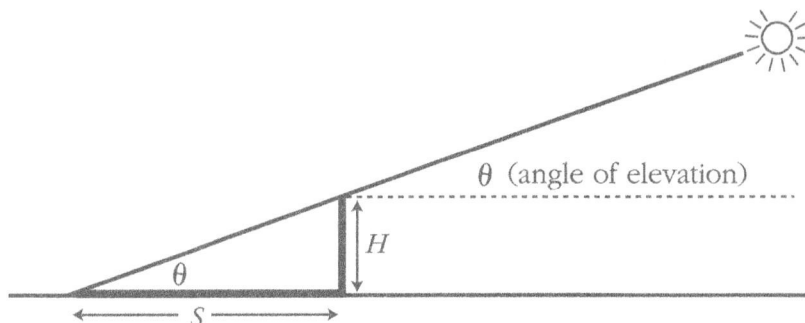

Beginning Portfolio Selection

Intent

Students begin assembling their portfolios, focusing on the central mathematical concept of this unit: similarity.

Mathematics

Students are asked to explain what similarity means by giving an intuitive description (same shape) and by using a formal definition (equal corresponding angles, proportional corresponding sides). Then they are asked to select two or three activities that helped them to understand this idea. The concept of similarity is introduced in *The Shape of It* and used throughout the rest of the unit.

Progression

Students review their notes and the text to locate activities that helped them understand and apply the concept of similarity.

Approximate Time

25 minutes (at home)

Classroom Organization

Individuals

Shadows Portfolio

Intent

Students compile their portfolios for the unit and write their cover letters. As this is the last unit of the year, an end-of-year reflection and review are also included.

Mathematics

In this portfolio, students will reflect on the mathematics covered in this unit and over the entire year.

Progression

Students start work on their portfolios in class by reading the instructions in the student book and beginning their cover letters. They complete their compilation of items and reflective writing outside of class.

Approximate Time

20 minutes for introduction

30 minutes for activity (at home)

Classroom Organization

Whole-class introduction, then individuals

Doing the Activity

Have students read the instructions in the student book carefully. Then they are to take out and review all of their work from the unit. They will have completed part of the selection process in *Beginning Portfolio Selection*. Their main task today is to write their cover letters.

Discussing and Debriefing the Activity

You may want to have students share their portfolios in their groups, comparing what they wrote about in their cover letters and the activities they selected.

1-Inch Graph Paper

$\frac{1}{4}$-Inch Graph Paper

1-Centimeter Graph Paper

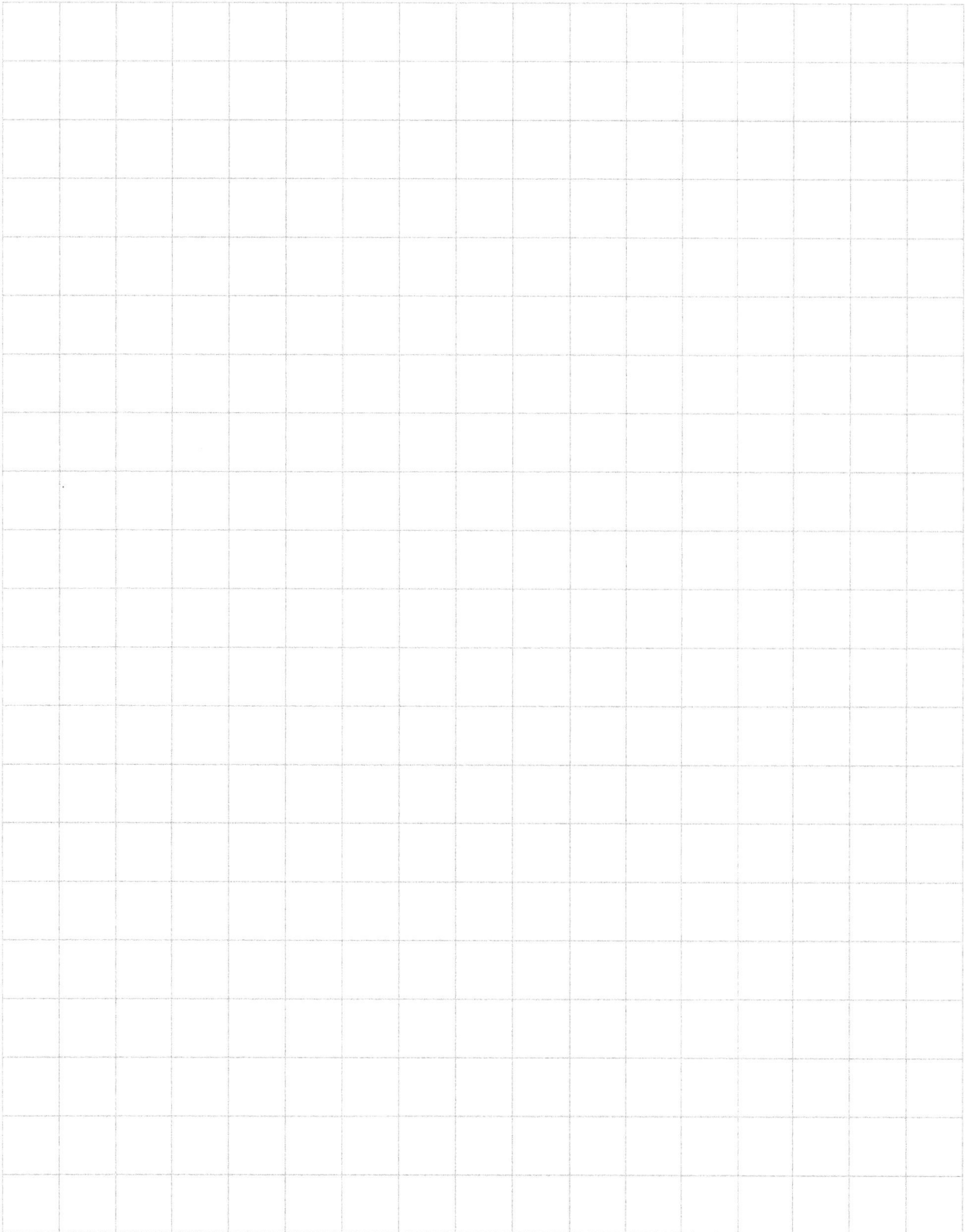

You place a tower of cubes 9 inches tall on a table. You shine a flashlight at the tower. The flashlight is mounted on a stand so that it is 24 inches above the tabletop.

The distance from the base of the tower to the spot on the table directly below the flashlight is 17 inches.

How long is the shadow cast by the tower? Explain your reasoning.

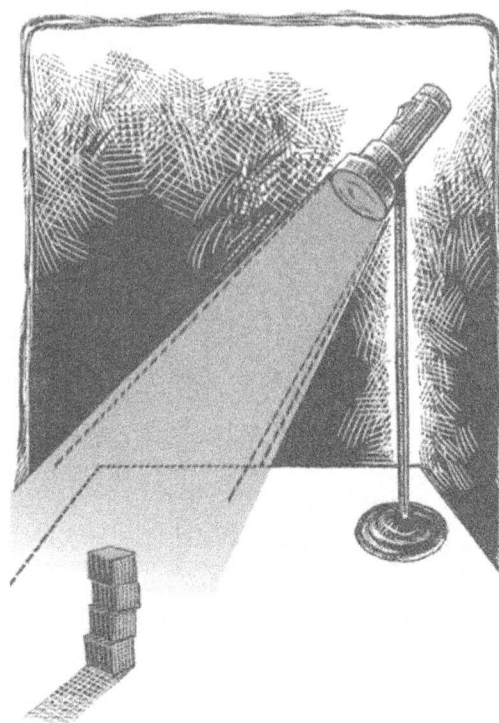

1. The Ladder

A ladder is leaning against a building.

The bottom of the ladder is 3 feet from the building. The ladder makes an angle of 75° with the ground.

Answer the following questions. Be sure to show your work.

a. How high up on the building does the ladder reach?

b. How long is the ladder?

2. Building Measurement

Find something in your neighborhood that is too tall for you to measure directly. For example, you might choose the height of your roof or the height of a tree.

a. Describe in detail *two ways* you could find the height of this object *indirectly*. Use ideas you learned in this unit. Be sure to explain why your methods work.

b. Carry out *one* of your plans. Give the specific measurements you make directly. Show how you use those measurements to find the object's height.

www.ingramcontent.com/pod-product-compliance
Lightning Source LLC
Chambersburg PA
CBHW051347200326
41521CB00014B/2508